T0140286

Studies in Computational Intelligence

Volume 698

Series editor

Janusz Kacprzyk, Polish Academy of Sciences, Warsaw, Poland
e-mail: kacprzyk@ibspan.waw.pl

About this Series

The series "Studies in Computational Intelligence" (SCI) publishes new developments and advances in the various areas of computational intelligence—quickly and with a high quality. The intent is to cover the theory, applications, and design methods of computational intelligence, as embedded in the fields of engineering, computer science, physics and life sciences, as well as the methodologies behind them. The series contains monographs, lecture notes and edited volumes in computational intelligence spanning the areas of neural networks, connectionist systems, genetic algorithms, evolutionary computation, artificial intelligence, cellular automata, self-organizing systems, soft computing, fuzzy systems, and hybrid intelligent systems. Of particular value to both the contributors and the readership are the short publication timeframe and the worldwide distribution, which enable both wide and rapid dissemination of research output.

More information about this series at http://www.springer.com/series/7092

Krishnanand N. Kaipa · Debasish Ghose

Glowworm Swarm Optimization

Theory, Algorithms, and Applications

 Springer

Krishnanand N. Kaipa
Department of Mechanical and Aerospace
   Engineering
Old Dominion University
Norfolk, VA
USA

Debasish Ghose
Department of Aerospace Engineering
Indian Institute of Science
Bangalore, Karnataka
India

This material is submitted for confidential review for possible book publication and should not be distributed for any other purposes

ISSN 1860-949X   ISSN 1860-9503   (electronic)
Studies in Computational Intelligence
ISBN 978-3-319-84694-1   ISBN 978-3-319-51595-3   (eBook)
DOI 10.1007/978-3-319-51595-3

Printed on acid-free paper

This Springer imprint is published by Springer Nature
The registered company is Springer International Publishing AG
The registered company address is: Gewerbestrasse 11, 6330 Cham, Switzerland

Preface

Stroll the meadows on a dark summer night and you may catch a glimpse of glowworms that illuminate the night with brilliant flashes of light, thereby making themselves conspicuously visible over long distances. Glowworms produce an emission of natural glow through a process called bioluminescence. They exploit the property of bioluminescence in a variety of ways for the purposes of mating and species preservation. Glowworms use photic signaling and attraction mechanisms to rendezvous on trees and form large swarms, resembling fiery clouds that flash synchronously. Firsthand accounts of this unique group behavior have been reported by many naturalists in the past.

The focus of this book is on the development of glowworm swarm optimization (GSO), a swarm intelligence algorithm inspired by the behavior of glowworms (also known as fireflies or lightning bugs). GSO belongs to a broad class of synthetic swarm algorithms that are modeled after natural swarms. The underlying designs either closely, or loosely, mimic the individual behaviors of their natural counterparts; in other frameworks, the mirrored principles may be modified to some extent based on the target problem. GSO is originally developed for numerical optimization problems that involve computing multiple optima of multimodal functions, as against other algorithms, which aim to identify the global optimum. The problem that GSO solves is inspired by swarm robotics applications involving search for possibly multiple unknown signal sources.

The behavior of natural glowworms to vary bioluminescence and glow at different intensities is replicated in GSO. Each synthetic glowworm glows with an intensity proportional to its fitness and moves toward a stochastically chosen neighbor with a relatively higher glow. An adaptive neighborhood range discounts the effect of distant members whenever a glowworm has a plenty of neighbors or the range goes beyond its perception limit. These taxis behaviors (based only on local information and selective neighbor interactions) enable the swarm to split into disjoint subgroups that converge to multiple high function value points.

The generality of the GSO algorithm can be evidenced in its application by independent researchers to a variety of problems ranging from optimization to robotics. Examples include computation of multiple optima, annual crop planning, cooperative exploration, distributed search, multiple source localization, contaminant boundary mapping, wireless sensor networks, clustering, knapsack, numerical integration, solving fixed point equations, solving systems of nonlinear equations, and engineering design optimization. Further, other researchers have also developed several variants of GSO in order to improve its convergence properties.

Overview of the Book

This book provides a comprehensive account of the glowworm swarm optimization algorithm. Various aspects of GSO including the underlying ideas, theoretical foundations, algorithm development, computer programs, possibly interesting research problems, and its variations are provided. The book is divided into eight chapters.

Chapter 1 provides an overview of natural and synthetic swarm intelligence. A brief account of bioluminescence in natural glowworm swarms is provided. The underlying ideas of the GSO algorithm and the behavior patterns that are borrowed from natural glowworms are described. The primary target problems solved by GSO are briefly described.

Chapter 2 presents the algorithm development of GSO. The various algorithm steps of the basic algorithm are described. GSO, in its present form, has evolved out of several significant modifications incorporated into the earlier versions of the algorithm. Many ideas were considered in the development process before converging upon the current GSO version. Some of the important steps in this evolution are briefly discussed. Simulations to illustrate the basic capability of the algorithm are presented. The various features of GSO are compared with those of other popular swarm intelligence algorithms. Additional notes on independent works on GSO are provided at the end of the chapter. A more in-depth treatment of this topic is deferred to Chap. 8.

Chapter 3 deals with the theoretical foundations of GSO. Initially, a characterization of the swarming behavior of agents in GSO is carried out, which guides the formulation of a framework used to analyze the algorithm. A theoretical model of GSO is obtained by using assumptions that make it amenable to analysis, yet reflecting most of the features of the original algorithm. Next, local convergence results for this GSO model are provided. Some illustrative simulations are presented to support these theoretical findings.

In Chap. 4, numerical simulation results to evaluate the efficacy of GSO in capturing multiple optima of multimodal optimization problems are presented. For this purpose, several benchmark multimodal functions are considered that pose a wide range of complexities. The parameter selection problem is addressed by conducting experiments to show that only two parameters need to be selected by the user. Next, simulation experiments are used to compare GSO with Niche-PSO, a PSO variant that is designed for the simultaneous computation of multiple optima. Finally, algorithmic behavior is examined in the presence of noise.

In Chap. 5, the potential of GSO for signal source localization is demonstrated by using physically realistic simulations and experiments with real robots. The modifications incorporated into the algorithm in order to make it suitable for a robotic implementation are described. Results from Player/Stage simulations and sound source localization and light source localization experiments are presented to test the potential of robots using GSO for localizing signal sources.

In Chap. 6, a GSO variant for a heterogeneous swarm of mobile and stationary agents is developed and its application to ubiquitous computing environments is discussed. In particular, a GSO-based ubiquitous computing environment is proposed to address the problem of sensing hazards. Simulation experiments are performed to demonstrate the efficacy of the algorithm in tackling such hazardous situations. It is shown that the deployment of stationary agents in a grid-configuration leads to multiple phase-transitions in a graph of minimum number of mobile agents required for 100.

In Chap. 7, the behavior of GSO agents in the presence of mobile sources is investigated. In particular, a coordination scheme based on GSO is developed that enables a swarm of glowworms to pursue a group of mobile signal sources. Some theoretical and numerical results that provide upper bounds on the relative speed of the mobile source in different cases are presented.

Chapter 8 presents a survey on applications of GSO and its extensions. Work on GSO that appeared in the recent literature can be primarily classified into three categories. Researchers in the first category proposed modifications of GSO, which were mainly focused on either improving the convergence properties of original GSO or modifying GSO for global optimization problems. In the second category, researchers used basic GSO in different applications. In the third category, other researchers modified GSO and used them in some applications. A summary of these independent works is provided.

Each chapter ends with a set of thought exercises and computer exercises. A long list of references on the GSO algorithm, along with its variants, and GSO applications are provided. Finally, GSO code in MATLAB and C++ and some useful external links are provided in the appendix.

The book is intended primarily for researchers in swarm intelligence and computational intelligence and graduate and undergraduate students working on these topics.

Norfolk, VA, USA   Krishnanand N. Kaipa
Bangalore, India   Debasish Ghose

Acknowledgements

Many people made valuable contributions to the topic of glowworm swarm optimization (GSO) and helped us to apply the technique to many problems in diverse fields ranging from optimization to robotics. Some of these are covered in this book.

Amruth Puttappa, Guruprasad M. Hegde, and Sharschandra V. Bidargaddi helped us in building some of the robotic platform used to test the feasibility of applying GSO to source localization tasks. Ruta Desai helped us to conduct GSO-based light localization experiments using the same robotic platform. Varun Raj Kompella designed and implemented embodied simulations of GSO using Player/Stage software, the results of which are reported in Chap. 5. Joseph Thomas was a Master's degree student who modified GSO and applied it to the problem of multiple odor source localization. In recent years, people who have collaborated with us on using or implementing GSO are Prathyush P. Menon, Achal Aravind, Ashish Derhgawen, Paloma Sodhi, and Ashish Budhiraja.

Thomas Ditzinger, senior editor of Springer Publishers, has coordinated with us since the time the first book's proposal was sent to Springer and has given us ample time to complete the final draft. The staff at Springer were very helpful in taking this book project to the completion stage.

Finally, we are thankful to our families. Krishna Kaipa would like to thank his father Bhaskar and mother Jhansi Rani for their extreme support when Krishna started the work on GSO at the Indian Institute of Science. Krishna would like to thank his wife Amy for her constant support as he worked on different portions of the book at University of Vermont, University of Maryland, and Old Dominion University. Krishna would also like to thank his golden retriever puppies Olli and Lumi for sitting by his side and giving him company for innumerable hours as he worked on the book project. Debasish would like to thank his wife Swati and son Debraj for being there and giving intangible but much-needed support in every possible way.

Contents

List of Figures

Chapter 1
From Natural to Synthetic Swarms

Nature abounds in examples of *swarming*, a form of collective behavior found in insect and animal societies. Glowworms use simple signaling and attraction mechanisms to congregate into large swarms for the purpose of mass mating. Ants use trail-laying and trail-following behaviors to self-organize into complex foraging patterns; ants gathered in groups can carry prey that are so large that if they were fragmented, the members of the group would be unable to carry all the fragments individually. Honeybee swarms use group decision making to find a future nest: some of the bees, called scout bees, find potential sites in all directions and advertise a dozen or more of them to recruit other bees to visit these sites, but eventually they reach a consensus about a single site. Schooling in fish serves to confuse a predator. Birds that gather in large flocks for migrations achieve aerodynamic efficiency, higher than that of a single bird, leading to reduced fatigue and higher chances of survival at the end of the migration. The decentralized mechanisms responsible for the swarm intelligence found in these examples, and others in the natural world, offer an insight into the basis to synthesize artificial techniques that can be applied to diverse fields such as optimization, multi-agent decision making, and collective robotics.

The focus of this book is glowworm swarm optimization (GSO), a swarm intelligence algorithm inspired by the behavior of glowworms (also known as fireflies or lightning bugs). GSO is originally developed for numerical optimization problems that involve computing multiple optima of multimodal functions, as against other swarm intelligence algorithms which aim to identify the global optimum. The problem that GSO solves is inspired by swarm robotics applications involving search for unknown signal sources. GSO belongs to a broad class of synthetic swarm algorithms that are modeled after natural swarms. These algorithms are usually based on designs in which the artificial agent protocols either closely, or loosely, mimic the individual behaviors of their natural counterparts; in other frameworks, the agent rules are borrowed from those of their real counterparts and modified to some extent based on the target problem. The rest of this chapter deals with a description of the decentralized mechanisms observed in natural swarms. Later, how computer scientists and

© Springer International Publishing AG 2017
K.N. Kaipa and D. Ghose, *Glowworm Swarm Optimization*,
Studies in Computational Intelligence 698, DOI 10.1007/978-3-319-51595-3_1

engineers translated some of these mechanisms into their synthetic counterparts that could solve problems in optimization and robotics, will be discussed.

1.1   Natural Swarm Intelligence

Swarm intelligence (SI), as observed in natural swarms, is the result of actions that individuals in the swarm perform exploiting local information, which is communicated, indirectly through the environment, by their neighbors [18]. One example of this swarming phenomenon that is well established and extensively studied is the group foraging by ants: An ant lays a chemical called pheromone on its way during its search for food, thus forming a pheromone trail; when a food source is found, it returns back to the nest, while simultaneously reinforcing the retraced trail with more pheromone. However, when another ant randomly searching for food encounters a pheromone trail by chance, it starts to follow the pheromone trail, with a high probability, and continues to lay its own pheromone, thereby further reinforcing the trail being followed; at a intersection of two trails, it selects to follow, with a higher probability, the trail with the higher pheromone level; eventually, it reaches either the food source or the nest. As the pheromones decay with time, pheromone trails that are circuitous and less visited by ants may be quickly lost in the due course of time; however, shorter routes between the food source and nest may be positively reinforced frequently enough before they are lost. This process eventually results in the emergence of a shortest route between the nest and the food source.

It can be seen in the above example how ants use pheromone trails collectively to exhibit swarm behavior. However, the method of communication and/or the purpose of swarming might differ from one species to another. For instance, honeybees use waggle dances performed by other honeybees as cues to find the direction and distance to food sites. Glowworms exploit the property of bioluminescence in a variety of ways for the purposes of mating and species preservation. As this book deals with an optimization algorithm modeled after the behavior of glowworms, the swarm intelligence principles observed in these luminous insects are discussed next.

1.1.1   Glowworm Swarms

Stroll into the meadows on a dark summer night and you may catch a glimpse of glowworms that illuminate the night with brilliant flashes of light, thereby, making themselves conspicuously visible over long distances. The term *glowworm*[1] com-

---

[1]The term *glowworm* is also used to describe:

1. the luminescent larvae of Mycetophilidae (order: *Diptera*) [124].
2. the luminescent larvae of certain fungus gnats that belong to the subfamilies *Arachnocampinae*, *Keroplatinae* and *Macrocerinae* of the dipteran family *Keroplatidae* [149].
3. the wingless adult females of Lampyridae (e.g., *Lampyrus noctiluca* of Europe) [16, 124].

monly refers to the larvae of fireflies (also known as lightning bugs) [8, 124]. Fireflies spend most of their lifetime as glowworms, that is, in the larval stage (two to three years versus a few weeks as an adult) [30]. Therefore, the terms adult glowworms, fireflies, and lightning bugs can be used interchangeably. In this book, they are just called glowworms, unless a different usage is required by the context.

Glowworms produce an emission of natural glow through a process called bioluminescence [194]. They belong to a family of beetles (Order: *Coleoptera*) called *Lampyridae*; together with *Phengodidae* and *Rhagophthalmidae*, they represent one of the few insect families in the whole of the animal kingdom that exhibit the power of emitting natural light [16]. However, this property is shared by other animals such as Protozoa (e.g., *Noctiluca*), Hydrozoa (e.g., *Pyrosoma*), several molluscs, and some vertebrates (e.g., some kind of fish). As a matter of fact, about 80 to 90 percent of the ocean life is composed of bioluminescent creatures [218], which also form about 80 percent of all the 700 genera known to contain luminous species [217].

An intermittent flashing of glowworms, typically found on open grounds (e.g., meadows, forest glades, and road clearings) of the West, is a very common sight [16]. However, a rather rare, and a strikingly remarkable, phenomenon is their complex swarming behavior: Hundreds of glowworms assemble on trees in large congregations and flash synchronously [21, 22, 70, 153, 198]. Several first-hand accounts of this unique group behavior have been reported in the literature (Refer to [70, 156] for a summary). Two of these accounts, based on the glowworm swarms found along the broad stretch of the Maenam Chao Phraya river in Thailand (formerly called Siam), are quoted here for a better visualization of this entrainment effect:

Engelbert Kaempfer, 1690 [70]: *The Glowworms (Cicindalae) represent another shew, which settle on some Trees, like a fiery cloud, with this surprising circumstance, that a whole swarm of these Insects, having taken possession of one Tree, and spread themselves over its branches, sometimes hide their Light all at once, and a moment after make it appear again with the utmost regularity and exactness, as if they were in a perpetual Systole and Diastole.*

H.M. Smith, 1935 [198]: *Imagine a tree thirty-five to forty feet high thickly covered with small ovate leaves, apparently with a firefly on every leaf and all the fireflies flashing in perfect unison at the rate of about three times in two seconds, the tree being in complete darkness between the flashes. Imagine a dozen such trees standing close together along the river's edge with synchronously flashing fireflies on every leaf. Imagine a tenth of a mile of river front with an unbroken line of Sonneratia trees with fireflies on every leaf flashing in synchronism, the insects on the trees at the ends of the line acting in perfect unison with those between. Then, if one's imagination is sufficiently vivid, he may form some conception of this amazing spectacle. By going out into the river far enough from shore to lose sight of the individual flashes, a person may obtain from a single tree, a group of trees or a long line of trees a weird pulsating mass effect.*

These interesting reports naturally lead us to the question of whether there exists any purpose underlying such a swarming behavior of glowworms. It is a fairly well established fact that glowworms use photic-signaling and photic-attraction, in a variety of ways, for mating purposes [16, 115]. Individual courtships are common among the glowworm species of the West. Examples include *Lampyrus noctiluca* (the common glowworm of Europe), *Luciola italica* (fireflies of southern Europe), and the American species of *Lampyridae* (popularly known as the lightning-bugs). However, the glowworms of the tropics have evolved mass mating [21], instead of pair courtship, the reasons for which are not very obvious. Also, it is not immediately clear whether or not the mass mating and the mass synchronous flashing are functionally related to each other.

The rest of the discussion about natural glowworms is devoted to an understanding of their photic communication and other physiological mechanisms that help us to answer some of the questions posed above. Another benefit of this endeavor is to extract principles that inspire the synthesis of artificial swarm systems.

Photic Communication

The average life expectancy of adult glowworms is only a few weeks [16, 30, 115]. During this short time, they are faced with an urgency to find a partner, mate, and reproduce for preservation of their species. To achieve this purpose, glowworms have evolved effective means of communicating mating signals by exploiting their ability to control the light emission in a variety of ways; they can generate a wide and diverse range of information signatures by modulating the following parameters of the glow: brightness, color, continuous glow or pulse train of flashes, number of flashes per cycle, duration of each flash, time period of flashes, and phase difference between male flashes and female flashes.

In *Lampyrus noctiluca*, the common glowworm of Europe, the property of glowing is confined only to the apterous female. She crawls over the grass, and sweeps her light[2] from one side to the other in order to attract the attention of wandering males. In *Lamprophorus tenebrosus,* the luminous beetle of Sri Lanka, both the male and the female are luminous. The female is wingless and exposes its light, much in a manner similar to Lampyrus, to call a mate. The male is normally brilliant. However, it shuts off its light while approaching a seeking female, who responds to its arrival by partially switching off her light. Unlike in the Lampyrus, both the sexes of the *Luciola italica* (the fireflies of southern Europe) are luminous. When seeking a mate the female glows with long flashes, incompletely extinguished during the intervals. This period of flashes is succeeded by a dark period. The males perceive such a pattern of flashing, from as far as ten feet away, pause their course of flight, and fly down to the female. Although males are attracted to some extent by a steady glow, they readily approach a female when she glows with a pattern of flashes.

---

[2]As the light organ is hung from underside of her tail, she twists her body to one side so as to display her light clearly.

Species-Specific Flashing Patterns

The lightning-bugs that belong to the genus *Photinus* have evolved different species-specific flashing patterns for species preservation through effective mate selection mechanisms; each species has a characteristic flashing pattern and an individual of any one species will reply to, or evoke a reply from, a member of the opposite sex of that species. McDermott [16] conducted experiments with small electric bulbs that simulated various flash patterns of this insect to confirm these findings.

Both sexes of Photnius are luminous and the light is emitted as a series of brilliant flashes. The male takes the initiative to search for a mate by hovering over the ground, flashing, and apparently watching for an answering flash from the female lying on the grass. Two or more species of Photinus can often be found on the same ground and at the same time. In *Photinus pyrails*, the male gives out a single flash. This is replied three or four seconds later, with a single flash of lower intensity and longer duration, by a female of the same species; this particular female was not found to reply to the flash by a male of another species (*Photinus consanguineus*). The male's recognition of the species-specific female flash pattern was confirmed when the male was observed to be attracted to a bulb placed in the grass that flashed three to four seconds after his own flash; the male did not show any interest when the bulb flashed without a pause. In *Photinus consanguineus*, the male gives out two flashes in quick succession, followed by a short pause, then two more, and so on. The female replies within a second to the second flash of the male. In *P. scintillans*, the male gives out a short single flash. The female replies with a longer single flash. She would also reply to the first flash of a male of *P. consanguineus*, but the latter doesn't take notice. In *P. marginellus*, the male gives out a single short sharp flash. The female replies quickly with a double flash, the first of which sharper and brighter than the second, followed at once by the second. In *P. castus*, the male gives out a single flash, not so short and sudden as *P. marginellus*. The female replies with a single flash, immediately after the flash of the male. Although *P. castus* and *P. marginellus* are very similar to each other in their structure, they are considered as distinct species based on the very different pattern of flashes emitted by them. They are frequently found flying together. However, no case of interbreeding was observed. This a very striking feature of glowworms, which is achieved by their ability to modulate, and recognize, a variety of flashing patterns. Next, we see how the glowworms of tropical swamplands evolved mass mating, instead of individual courtships.

Mass Mating and Mass Synchronous Flashing

In all the species mentioned above, as those typically found in the meadows of the West, the males hover above the ground and search for stationary females. Recognition of, taxis toward, and landing next to, the matching partner depend on the availability of an uninterrupted line-of-sight vision in these regions, which allows them to distinguish the flash signature of their own species from those of other species patrolling the same terrain. However, the visually cluttered mangrove swamps of southeast Asia are not favorable for such a sustained photic communication between individuals. Therefore, the system of individual courtships is not a mating option for the glowworm species found here. A promising hypothesis was proposed by Buck and Buck [21], based on their observations of the Thailand *Pteroptyx*: a tree on a watercourse hosting a swarm of illuminated glowworms might act as a sufficiently bright and large beacon to attract other glowworms that wander out into the clear over the water, and provide enough mating opportunities to compensate for the long flights required of the mated females for egg dispersal. If no previous swarm existed, the wandering glowworms might themselves form a nucleus by mutual photic attraction. Such congregations build up competitively, leading eventually to one or a few large swarms that outdraw nearby smaller centers because of higher mean light emission that is indicative of a higher likelihood of finding a mate.

Some more examples of swarming behavior—the group foraging in ants, division of labor in honeybees, predator evasion in fish schools, formation flight in bird flocks—found in nature, are discussed next.

1.1.2 Ant Colonies

As mentioned earlier, ants gathered in groups are superefficient in the sense that they can carry prey items that are so large that if they were fragmented, the members of the group individually would be unable to carry all the fragments [54]. Colonies of the army ant *Eciton burchelli* stage huge swarm raids, in pursuit of arthropod preys, where up to 2,00,000 foragers transport more than 3000 prey items per hour over raiding columns that exceed 100 meters in length [55]. Couzin and Franks [34] show how the movement rules of these army ants on trails can lead to a collective choice of direction and the formation of distinct traffic lanes that minimize congestion. Deneubourg et al. [40] develop a minimal mathematical model based on a binary bridge experiment to show that workers of the Argentine ant use simple trail-laying and trail-following behaviors to self-organize into complex foraging patterns. Bernstein [14] studied a certain species of desert ants and showed that individuals optimize energetic efficiency by choice of foraging strategies. In particular, foragers searched for food individually, without apparent behavioral cohesion, under conditions of high food density. However, when the food density was relatively low, colonies switched to group foraging along defined trails/columns.

1.1.3 Honeybee Swarms

A honeybee that returns to the nest from a food source performs a waggle dance to signal its nest mates about the direction and distance to the forage site visited by the dancing bee [210]. The information encoded in the symbolic language is then used by unemployed foragers to locate the forage site. The proportion of scouts (foragers that search for new food sources independently) and recruits (foragers that use information cues from scouts performing recruitment dances, to guide their movements toward food sources) varies between about 5 and 35%, depending on forage availability [190]. Beekman et al. [13] developed a mathematical model to demonstrate that the decision to become a scout or a recruit could be regulated by whether a potential forager can find a recruitment dance within a certain time period. Seeley and Buhrman [189] analyze group decision making in honeybee swarms in the context of finding future home sites by honeybees. Their studies lead to some interesting findings at the group level of the swarm: the scout bees find potential nest sites in all directions and at distances of up to several kilometers. Initially, the scouts advertise a dozen or more sites with their dances on the swarm, but eventually they advertise just one site. Within about an hour of the appearance of consensus among the dancers, there is a crescendo of dancing and later the swarm lifts off to fly to the chosen site. However, the chosen site is not necessarily the one that is first advertised on the swarm.

1.1.4 Fish Schools and Shoals

Schooling in fish serves as an anti-predator function by confusing the predator [57]. Apparent predation risk and food availability result in changes in the inter-individual distances that fish maintain. For example, fish increase nearest-neighbor distances when hungry leading to groups of smaller, dispersed, and less compact shoals [155]. However, they position themselves closer to other individuals, leading to one compact shoal of large group size,[3] when attacked by a predator [130]. Hoare et al. [80] experimentally examine context-dependent group size choice in a shoaling fish, the banded killifish, *Fundulus diaphanus*, by using nondirectional odor cues to simulate a food source or a successful attack by a predator. They show that the group sizes are significantly smaller in the food treatment and larger in the alarm treatment than in control trials where the food or alarm treatments are absent.

1.1.5 Bird Flocks

Birds that gather in large flocks for their long semiannual migrations enjoy several advantages over an isolated individual [68]. Birds in some specific line formations, with Geese as the common example, can achieve group aerodynamic efficiency higher than that of a single bird [122]. The resulting decrease in energy consumption and fatigue leads to a higher chance of survival at the end of the migration. However, Higdon and Corrsin [79] note that many species like Passerines migrate in large

---

[3]Here the group size refers to the number of fish in the group and should not be confused with the size of the area occupied by the group.

three-dimensional flocks; they develop a mathematical model based on simple classical steady-state aerodynamic theory to estimate the three-dimensional effects for simple lattice formations and show that the total drag decreases when the flock extends farther laterally than vertically.

1.2 Synthetic Swarm Intelligence

The swarming mechanisms found in the examples discussed in the previous section, and others in the natural world, offer an insight into the basis to devise decentralized decision-making algorithms that solve complex problems related to diverse fields such as optimization, multi-agent decision making, and collective robotics. Recent literature abounds with examples of such swarm-inspired algorithms including ant colony optimization (ACO) techniques [15, 18, 19, 44, 48, 49], particle swarm optimization (PSO) algorithms [20, 29, 97, 98, 117, 126, 167, 168], and several collective robotics algorithms [36, 73, 104, 105, 127, 227].

1.2.1 Ant Colony Optimization

Ant colony optimization (ACO) was introduced by Marco Dorigo in 1992 [44]. It is inspired by the foraging behavior of ant colonies and represents one of the most successful examples of synthetic swarm intelligence algorithms. ACO was originally developed to solve discrete optimization problems. The classic traveling salesman problem (TSP), a NP-hard optimization problem, is a typical application of ACO in which the pheromone update on, and the following of, city routes closely mimics the trail-laying and trail-following behaviors of real ants [19]. The TSP consists of finding the shortest tour between n cities visiting each once only and ending at the starting point. The TSP application of ACO starts by each artificial ant visiting the n cities to construct a tour. After completing its tour it lays artificial pheromone on the used links with an amount that is proportional to the quality of the tour. As a result of this process, links which belong to good solutions end up with more pheromone than the other links. During the construction of a tour, each ant at a current city tends to select the next city that is connected by a link with more pheromone than the others. This selection process amplifies previously reinforced links and leads to the emergence of a good solution. Pheromone decays at a fixed rate after all ants have constructed their tours; this pheromonal decay introduces a forgetting factor that prevents inferior links from being amplified by accident.

The original ACO algorithm is known as the Ant System (AS) [44, 47]. The AS algorithm is initialized by placing m ants on different cities randomly and by assigning initial values of pheromone $\tau_{ij}(0)$ to each edge (i, j) joining any two cities i and j. Next, the algorithm executes one instance of a cycle comprising a solution construction phase that is followed by a pheromone update phase.

During the solution construction phase, each ant k, starting from its initial city, builds a tour by visiting each of the remaining $(n-1)$ cities exactly once. The ants select the next city to be visited using a stochastic mechanism. In particular, when an ant k, having constructed a partial tour s^P so far, is located at city i, the probability of visiting city j is given by:

$$p_{ij}^k = \begin{cases} \frac{\tau_{ij}^\alpha \cdot \eta_{ij}^\beta}{\sum_{c_{il} \in N(s^P)} \tau_{il}^\alpha \cdot \eta_{il}^\beta} & \text{if} c_{ij} \in N(s^P), \\ 0 & \text{otherwise,} \end{cases} \tag{1.1}$$

where, $N(s^P)$ is the set of feasible components and consists of edges (i, l) where l is a city not yet visited by the ant k. The parameters α and β control the relative importance of the pheromone versus the heuristic information η_{ij}, which is given by:

$$\eta_{ij} = \frac{1}{d_{ij}}, \tag{1.2}$$

where, d_{ij} is the distance between the cities i and j.

During the pheromone update phase, the pheromone values, associated with the edges joining the cities, are updated by all the ants that have built a solution within that iteration. Accordingly, the pheromone τ_{ij}, associated with the edge joining cities i and j, is updated as follows:

$$\tau_{ij} \leftarrow (1 - \rho) \cdot \tau_{ij} + \sum_{i=1}^{m} \Delta \tau_{ij}^k \tag{1.3}$$

where, ρ is the evaporation rate, m is the number of ants, and τ_{ij}^k is the quantity of pheromone laid on edge (i, j) by ant k:

$$\Delta \tau_{ij} = \begin{cases} Q/L_k & \text{if ant } k \text{ used edge}(i, j) \text{ in its tour,} \\ 0 & \text{otherwise,} \end{cases} \tag{1.4}$$

where, Q is a constant and L_k is the length of the tour constructed by ant k.

At this point, the shortest path found by the ants is saved, and the above cycle is repeated until a termination condition is satisfied: either the maximum number of cycles (user-defined) is reached or all the ants make the same tour.

After AS, several ACO variants have been proposed till now [45]. The two most successful variants are the MAX-MIN Ant System (MMAS) [201] and the Ant Colony System (ACS) [46]. The MMAS differs from the original AS in that only the best ant updates the pheromone trails and that the value of the pheromone is bounded from above and below. In ACS, an additional local pheromone update is introduced. In particular, after each construction step, the pheromone of the last edge traversed by each ant is updated as follows:

$$\tau_{ij} = (1 - \psi)\tau_{ij} + \psi\tau_0 \qquad\qquad (1.5)$$

where, $\psi \in (0, 1]$ is the pheromone decay constant, and τ_0 is the initial value of the pheromone.

Note that the effect of the local pheromone update given by (1.5) is to decrease the pheromone concentration on the traversed edges. This encourages subsequent ants to choose other edges and diversify their search, thereby decreasing the likelihood of several ants producing identical solutions during one iteration.

Examples of other successful ACO variants include Elitist AS [44, 47], Ant-Q [58], Rank-based AS [23], ANTS [134], and Hypercube-AS [17].

ACO can be used to quickly find high quality solutions to NP-Hard problems. Consequently, ACO has been tested on more than 100 different NP-hard problems. These problems can be categorized as follows:

1. Routing problems arising during distribution of goods (e.g., Traveling salesman, vehicle routing, and sequential ordering).
2. Assignment problems, where a set of items have to be assigned to a set of resources subject to some constraints (e.g., Quadratic assignment, graph coloring, and course timetabling).
3. Scheduling problems, where resources have to be allocated to tasks over time (e.g., Open shop, total tardiness, total weighted tardiness, and project scheduling).
4. Subset problems, where a solution to a problem is considered to be a subset of available items (e.g., Multiple knapsack, set covering, 1-cardinality trees, and maximum clique).

In many of the above problems, the performance of ACO algorithms was shown to be very close to those of the best-performing algorithms. On problems including sequential ordering, open shop scheduling, some variants of vehicle routing, classification, and protein-ligand docking, ACO was found to be the state-of-the-art. ACO algorithms are very effective for routing problems in telecommunication networks. For example, AntNet [24] is the state-of-the-art ACO algorithm that provides highly adaptive and robust strategies for communication in wired networks. Recently, there have been some attempts on designing ACO for continuous optimization [15, 199]. Research in this direction is currently ongoing.

1.2.2   Particle Swarm Optimization

Particle swarm optimization (PSO) was introduced by Kennedy and Eberhart in 1995 [98]. It is a population-based stochastic search algorithm inspired by the flocking behavior of birds and developed for numerical optimization. In PSO, a population of solutions $\{X_i : X_i \in R^m, i = 1, \ldots, N\}$ is evolved, which is modeled by a swarm of particles that start from random positions in the objective function space and move through it searching for optima; the velocity V_i of each particle i is dynamically adjusted according to the *best so far* positions visited by itself (P_i) and the whole swarm (P_g). The position and velocity updates of the ith particle are given by:

$$V_i(t+1) = V_i(t) + c_1 r_1 (P_i(t) - X_i(t)) + c_2 r_2 (P_g(t) - X_i(t)) \qquad (1.6)$$
$$X_i(t+1) = X_i(t) + V_i(t+1) \qquad (1.7)$$

where, c_1 and c_2 are positive constants, referred to as the *cognitive* and *social* parameters, respectively; r_1 and r_2 are random numbers that are uniformly distributed within the interval $[0, 1]$; and $t = 1, 2, \ldots,$ indicates the iteration number.

Many variants of PSO have been developed and tested since the introduction of the basic PSO. Several researchers have modified PSO to allow it to operate in binary spaces. For instance, Kennedy and Eberhart [99] described a binary particle swarm, in which the velocity is used as a probability threshold to determine whether the position bit should be evaluated as zero or one. Several techniques based on PSO have been proposed to tackle dynamic problems that are modeled by time-varying fitness functions [168]. Recently, PSO variants such as species-based PSO [117] and NichePSO [20] have been developed for computation of multiple local optima of multimodal functions. The interested reader is referred to a comprehensive review of PSO techniques in [176] and PSO applications in [175].

1.2.3 The Bees Algorithm

The bees algorithm is inspired by the foraging behavior of honeybee swarms and can be used mainly for combinatorial and functional optimization [173]. Scouting and differential recruitment are the key principles underlying the bees algorithm. The algorithm starts by placing the *scout* bees randomly in the search space and evaluating the fitnesses of the visited sites. The sites visited by the elite bees—the bees having the highest fitnesses—are chosen for neighbourhood search. The fitness value of each site is used to determine the probability of the bees being selected for that site. Therefore, the algorithm conducts searches in the neighborhood of the selected sites by assigning more bees to search near the best elite sites than those to search in the neighborhood of remaining sites. For each patch only the bee with the highest fitness will be selected to form the next bee population. The remaining bees in the population are assigned randomly around the search space looking for new potential solutions. These steps are repeated until a stopping criterion is met.

1.3 Glowworm Swarm Optimization

This book presents glowworm swarm optimization (GSO), a swarm intelligence algorithm modeled after the behavior of glowworms (also known as fireflies or lightning bugs). GSO was introduced by Krishnanand and Ghose in 2005 [105]. Subsequently, the authors used GSO in various applications and several papers on this work have appeared in the literature [103, 104, 106–110, 112, 113]. Later, other researchers have proposed the firefly algorithm [222], which essentially follows the same logic as GSO with some minor variations. GSO is originally developed for numerical optimization problems that involve computing multiple optima of multimodal functions, as against other swarm intelligence algorithms which aim to identify the global optimum. The problem that GSO solves is inspired by However, the algorithm also prescribes decentralized decision-making and movement protocols useful for swarm robotics applications. For instance, a robot swarm can use GSO to carry out disaster response tasks comprising search for multiple unknown signal sources; examples of such sources include nuclear spills, hazardous chemical spills, leaks in pressurized systems, and fire-origins in forest fires.

It has been recognized that GSO's approach of explicitly addressing the issue of partitioning a swarm required by multiple source localization is very effective [143]. The authors also believe that GSO is the first algorithm to do this and prior methods provided only indirect solutions to the problem.

1.3.1   Basic Principle of GSO

GSO is developed based on the behavior of glowworms. The behavior pattern of glowworms which is used for this algorithm is their apparent capability to change the intensity of bioluminescence and thus appear to glow at different intensities. Each glowworm or agent in GSO is assumed to carry a luminous pigment called *luciferin*, whose quantity encodes the fitness of its location in the objective space. This allows the agent to glow at an intensity approximately proportional to the function value being optimized. It is assumed that agents glowing with brighter intensities attract those glowing with lower intensity. In particular, each glowworm selects, using a probabilistic mechanism, a neighbor that has a luciferin value higher than its own and moves toward it. The algorithm incorporates an adaptive neighborhood range by which the effect of distant glowworms are discounted when a glowworm has sufficient number of neighbors or the range goes beyond their maximum range of perception. These movements—based only on local information and selective neighbor interactions—enable the swarm of glowworms to split into disjoint subgroups that converge to high function value points. This property of the algorithm allows it to be used to identify multiple optima of a multimodal function.

Natural glowworms perform individual courtships by using the bioluminescent light to signal, and taxis toward, other individuals of the same species. In the case of mass mating, they initially gather to form a few nuclei by mutual photic attraction. Later, these nuclei of glowworms act as bright beacons to attract others from far away distances, eventually forming large congregations. In these congregations, centers of higher mean intensity outdraw nearby ones of lower mean intensity (Refer to Sect. 1.1 for more details). The general idea in GSO is similar in the following two aspects:

1. Agents are assumed to be attracted to move toward other agents that have higher luciferin value (brighter luminescence).
2. The multiple peaks can be likened to the nuclei of glowworms, formed during the beginning of the mass mating process, that serve as bright beacons.

1.3.2 Multimodal Function Optimization

The traditional class of problems related to this topic focused on developing algorithms to find either one or all the global optima of the given multimodal function, while avoiding local optima. However, there is another class of optimization problems, which is different from the problem of finding only the global optima. The objective of this latter class is to find multiple local optima [20, 64, 72, 83, 97, 117, 126, 131, 132, 146, 151, 172, 195]. The knowledge of multiple local and global optima has several advantages such as obtaining an insight into the function landscape and selecting an alternative solution when the dynamic nature of constraints in the search space makes a previous optimum solution infeasible to implement. One of the domains that requires identification of multiple optima includes the learning of useful financial decision making from examples, where identifying relevant knowledge and useful intermediate concepts is important [195]. Another example is to identify a set of diverse rules that together can be used as the basis for a classifier [83]. Multi-modality in a search and optimization problem gives rise to several attractors and thereby presents a challenge to any optimization algorithm in terms of finding global optimum solutions. However, the problem is compounded when multiple (global and local) optima are sought.

Evolutionary computation (EC) and swarm intelligence (SI) represent two primary population-based approaches that are particularly suited to solving multimodal function optimization problems. Seeking multiple optima by maintaining population diversity has received some attention in the domain of genetic algorithms [64, 72, 83, 131, 132, 146, 151, 172, 195] and particle swarm optimization algorithms [20, 97, 117, 126]. Traditional genetic algorithms (GAs) can successfully identify the best rule (optimum) in the domain, but are incapable of maintaining rules of secondary importance [151]. Several niche-preserving techniques have been proposed that allow a GA to identify multiple optima of a multimodal function [64, 72, 131, 132, 146, 172]. The niche metaphor is inspired from nature where different subspaces (niches) within an environment support different types of species. The number of organisms contained within a niche is determined by the carrying capacity of the niche and the efficiency of each organism at exploiting niche fertility.

In niching methods, each peak of a multimodal domain is thought of as a niche that can support a certain number of concepts. The number of individuals supported by a niche is in direct proportion to the niche's carrying capacity, as measured by the niche peak's fitness relative to other niche peaks' fitness values present in the domain. As the number of individuals contained within a niche indirectly indicates the amount of computational effort the GA will spend to improve the niche, the niches are populated according to their fitness relative to the other peaks. In this manner, genetic algorithms can maintain the population diversity of its members in a multimodal domain. Some of the niching schemes that allow a GA to capture multiple optima of multimodal functions include sharing [64], clearing [172], deterministic and probabilistic crowding [131], [146], and restricted tournament selection [72].

Parsopoulos et al. [167] studied altering the fitness value via fitness function stretching to adapt PSO to sequentially find peaks in a multimodal environment. In particular, a potentially good solution is isolated once it is found (if its fitness is below a threshold value) and then the fitness landscape is stretched to keep other particles away from this area of the search space. The isolated particle is checked to see if it is a global optimum, and if it is below the desired accuracy, a small population is generated around this particle to allow a finer search in this area. The main swarm continues its search for the rest of the search space for other potential global optima. With this modification, the algorithm was able to locate all the global optima of the test functions successfully.

Brits et al. [20] proposed a niching particle swarm optimization (NichePSO) algorithm, which had some improvements over the model proposed in [167]. The NichePSO is a parallel niching algorithm that locates and tracks multiple solutions simultaneously. Løvbjerg et al. [126] use subswarms to improve swarm diversity and avoid premature convergence. In NichePSO, this approach is adapted to maintain and optimize niches in the objective function space. The success of the NichePSO depends on the proper initial distribution of particles throughout the search space. The authors use Faure-sequences [204] to ensure uniform distribution of initial particle positions. Particles in the main swarm do not share knowledge about the best solution and use only their own knowledge. If a particle's fitness shows very little change over a small number of iterations, a subswarm is created with the particle and its closest topological neighbor. A subswarm radius is defined as the maximum distance between the global best particle (particle with the maximum fitness in the subswarm) and any other particle within the subswarm. A particle that is outside of the swarm radius of a subswarm is merged into it when the particle moves into the swarm, that is, when its distance to the global best particle becomes less than the swarm radius. In this manner, population diversity is preserved in the form of different niches (subswarms) and each subswarm converges to a different optimum solution. The algorithm was reported to be successful at detecting global maxima and, sometimes, local maxima.

Kennedy [97] investigated modifying the PSO algorithm with stereotyping where clustering based on a particle's previous position, with cluster centers substituted for individual's or neighbor's previous bests, forces the clusters to focus on local regions.

Li [117] proposed a species-based PSO (SPSO) that incorporates the idea of species into PSO for solving multimodal optimization problems. Initially, a population of particles is generated randomly. Next, multiple species are formed and a species seed is identified for each species. For this purpose, all the particles are evaluated and sorted in descending order of their fitness values. The particle with the best fitness is set as the initial species seed. All particles that are within a radius r_s

of the species seed's position, along with the seed, form one species. The next best particle that falls outside the r_s range of the first seed is set as the next species seed. The above process is repeated until all the particles are checked against the species seeds. These multiple adaptively formed species are then used to optimize toward multiple optima in parallel, without interference across different species.

In [168], Parsopoulos and Vrahatis combined the constriction-factor based PSO variant with a repulsion technique in order to detect multiple global optimizers in a sequential manner. In particular, after the detection of a minimizer, in addition to the application of stretching [167], each particle is checked whether it lies within the repulsion area, whose range is determined by a user defined parameter, of the minimizer. If the particle lies within the bounds, then it is repelled away from the center of the repulsion region.

1.3.3 Multiple Source Localization

Localization of signal sources using mobile robot swarms has received some attention recently in the collective robotics community. Examples of such signal-sources include sound, heat, light, leaks in pressurized systems [56], hazardous plumes/aerosols resulting from nuclear/chemical spills [227], fire-origins in forest fires [25], deep-sea hydrothermal vent plumes [88], hazardous chemical discharge in water bodies [53], oil spills [28], etc. The problem is compounded when there are multiple sources. For instance, several forest fires at different locations give rise to a temperature profile that peaks at the locations of the fire. Similar phenomenon can be observed in nuclear radiations and electromagnetic radiations from signal sources. In all the above situations, there is an imperative need to simultaneously identify and neutralize all the sources using a swarm of robots before they cause a great loss to the environment and people in the vicinity.

Based on the nature of the emission source and the ambient medium, the source localization problem can be broadly classified into two categories:

1. Signals such as sound, light, and other electromagnetic radiations propagate in the form of a wave [7]. Therefore, the nominal source profile that spreads in the environment can be represented as a multimodal function and hence, the problem of localizing their respective origins can be modeled as optimization of multimodal functions.

2. Chemical signals that are emitted by sources such as hazardous chemical aerosols, gas leaks, etc., disperse through the environment by molecular diffusion and bulk flow [7, 161]. The chemical source gradually dissolves into the ambient fluid medium resulting in odor plumes. A plume can be defined as those regions of space that contain the set of all molecules released from a single source [66]. Turbulence in the flow of ambient fluid renders the plume discontinuous and

patchy where patches of ambient fluid are interposed between patches of odor [160]. Owing to reasons mentioned above, the source profile in these cases cannot be modeled using a static multimodal function and should be represented using dynamic gas models that account for factors like diffusion and turbulence.

The basic GSO algorithm is suited to the first category of source localization problems, in which the assumption of using a multimodal function to represent the nominal source profile is justified. Therefore, in this book, GSO is applied to the first category of problems. However, Thomas and Ghose [206] modified GSO and applied it to the second category of problems.

While the application of evolutionary approaches is largely limited to numerical (discrete combinatorial or continuous) optimization problems that involve computer-based experiments, the particle-nature of individuals in swarm based optimization algorithms enable their application to realistic collective robotics tasks, with some modifications where necessary. Swarm robotics is increasingly being used for source detection applications [36, 73, 104, 105, 127, 185, 227]. Other approaches to robotic signal source localization include occupancy grid-based [88], gradient-based [185], and model-based [53] strategies.

Sandini et al. [185] present a robotic plume tracing strategy where the robot computes temporally integrated differential gradients by using the instantaneous concentration measurements obtained from a pair of sensors and uses such gradient information to reach the point of highest concentration of a gas leak. These authors also propose a cooperative strategy where robots broadcast their sensed concentration within the environment such that each robot's movements is driven not only by the self-measure of concentration, but also by the indirect measure of concentration received from neighbors.

Hayes et al. [73] describe a spiral-surge algorithm in which a collection of autonomous mobile robots use spiral plume finding, surge, and spiral casting behaviors to find the source of an odor plume. The authors validate the algorithm on a group of real robots and embodied simulations.

Lytridis et al. [127] report the results of using cooperation among a group of mobile robots that implement a combination of chemotaxis and biased random walk (BRW) strategies to perform odor source localization tasks. The robots exchange information about their current position and the field strength at their current positions, thereby enabling a robot, that detects a weaker field, to modify its heading toward others, which detect a stronger field, by an amount proportional to the difference of the two field strengths.

Cui et al. [36] present a multiple source localization algorithm in which mobile agents use a grid map to represent an unknown environment, collect concentration values from all other agents through ad-hoc communication, and calculate a positive gradient direction using a biasing expansion swarm approach (BESA). According to this approach, each agent moves to one of the neighboring eight cells on which the net influence (called the biasing parameter) of other agents is maximum. Each agent's influence is proportional to the concentration of the diffused source at its location and inversely proportional to the square of its distance to the cell. The cohesion property of this swarming approach maintains the connectivity of the ad-hoc network. Note that each agent uses global information to decide its movements. The authors validate their approach only in non-embodied simulations. Even though the authors consider multiple sources, the algorithm does not explicitly take into account the interference problems that are caused due to multiplicity of sources.

Zarzhitsky et al. [227] design a swarm robotic network based on the principles of physicomimetics and fluid physics for chemical plume tracing tasks (CPT). Their approach is based on the notion of volumetric control that involves monitoring and responding to environmental conditions within a specified physical region. First, a physicomimetics, or artificial physics (AP) framework is used to assemble the robots into a lattice formation. Second, the robot lattice uses a framework based on computational fluid dynamics to guide its movements toward the plume source. In particular, it performs *fluxotaxis* by computing the local divergence of mass flux, and then following its gradient. If the robot lattice surrounds a suspected plume source, and the total mass flux exiting the circle of robots consistently exceeds a small threshold, which is empirically determined, then the robots have localized the true source. The physicomimetics obeys Newtonian physics and is amenable to Newtonian analysis. The chemical plume tracing approach is likewise analyzable using fluid physics. Therefore, the authors claim that their approach to the CPT problem can be characterized as theoretically sound. The authors use computer based simulations to validate their methods. The authors show that their approach of fluxotaxis outperforms two most popular techniques, chemotaxis and anemotaxis, over a large set of different flows, plumes, and tracing environments.

1.3.4 GSO for Multi-agent Rendezvous

Consensus problems in multi-agent networks, in particular, robotic networks, appear in various forms [6, 200, 209], where several agents transit from an initially random state to a final steady state in which all the members of the group agree upon their state values. The agent's state could represent physical quantities such as heading angle, position, frequency of oscillation, and so on. Vicsek et al. [209] analyze alignment of heading angles of multiple particles using the approach of statistical mechanics. In synchronization of coupled oscillators, a consensus is reached regarding the frequency of oscillation of all agents [200]. The multi-agent rendezvous problem, posed by Ando et al. [6], involves devising local control laws that enable all the members to steer toward, and eventually meet at, a single unspecified location. A variation of this problem may require subgroups of mobile agents to converge at different locations. GSO prescribes agent protocols to solve a closely related problem in which the rendezvous locations are assigned to the multiple signal centers that are unknown *a priori*.

1.4 Summary

In this chapter, examples of natural swarm intelligence principles observed in some insect and animal societies were described; special emphasis was given to the description about photic signalling and photic attraction behaviors in glowworm swarms. This was followed by details on how these notions have been exploited by computer scientists and engineers to design synthetic swarm techniques such as ant colony optimization, particle swarm optimization, honey bee optimization, and other decentralized algorithms serving collective robotics tasks. Later, glowworm swarm optimization (GSO), a novel swarm intelligence based algorithm for the simultaneous computation of multiple optima of multimodal functions was introduced. Previous work on the key applications of GSO, multimodal function optimization and multiple signal source localization, were discussed. Finally, the problem of multi-agent rendezvous and how GSO can be used as a protocol to solve a variation of this problem was briefly discussed.

1.5 Thought and Computer Exercises

Note: Solutions to the exercises in this chapter will aid in obtaining an insight into the underlying philosophy of GSO. A more formal understanding along with algorithmic details will become clearer in Chap. 2.

Exercise 1.1 What aspects of natural glowworms are copied into the properties of synthetic glowworms in GSO? What other behavioral mechanisms are used by the synthetic glowworms that are not found in their natural counterparts? Can you point out any commonalities and/or differences in their swarming behavior? (You may like to look up other sources in biology in order to come up with insightful answers to these questions. Answering them will be useful in understanding the algorithmic development of GSO introduced in Chap. 2).

Exercise 1.2 Describe the mechanism that enables the glowworm swarm to encode the fitness landscape of a multimodal function. How is a neighbor defined in GSO? What is the communication protocol used by the glowworms to share fitness information between neighbors? Discuss the scalability benefits offered by this protocol.

Exercise 1.3 Define the problem of multimodal function optimization in the context of GSO. How is it different from the goals of traditional optimization? Elucidate the relevance of this problem by using a few application scenarios. What are the other methods that can be used to solve this class of optimization problems? How does GSO differ from these methods?

Exercise 1.4 Spontaneous local interactions, self-organization, and emergent group behavior are some of the primary properties of a swarm intelligent system. Explain how these properties are achieved by the GSO algorithm. What is the causal role played by the adaptive neighborhood in the emergence of swarm-splitting patterns in GSO? Explain how this behavior pattern helps in finding multiple optima of multimodal functions?

Exercise 1.5 Define the problem of multiple source localization. Give some examples. Discuss how a group of robots can use GSO to carry out source localization tasks (Hint: GSO was developed to solve numerical optimization problems. Is it possible to utilize it directly for a robotic implementation? If not, what modifications are necessary? Refer to a survey by McGill and Taylor [143] on robot algorithms for multiple emission source localization applications for related information on this topic).

Exercise 1.6 The general notion of multi-agent rendezvous involves multiple agents that start from different initial states, undergo state transitions, and eventually end up in the same final state. How does this phenomenon manifest in source localization tasks? (Hint: What do agent and state translate to in these scenarios?) How can this formulation be extended to the problem of multiple source localization?

Chapter 2
Glowworm Swarm Optimization: Algorithm Development

In this chapter, the development of the glowworm swarm optimization (GSO) algorithm is presented. Initially, the basic working principle of GSO is introduced, which is followed by a description of the phases that constitute each cycle of the algorithm. GSO, in its present form, has evolved out of several significant modifications incorporated into the earlier versions of the algorithm. Many ideas were considered in the development process before converging upon the current GSO version. Some of the important steps in this evolution are briefly discussed. Next, some convergence proofs related to the luciferin update rule of GSO are provided. Later, simulations are used to illustrate the basic working of GSO. Finally, some comparisons of GSO with ACO and PSO are provided based on their underlying principles, algorithmic aspects, and applications. Experimental comparisons of GSO with a variant of PSO are deferred to Chap. 4.

2.1 The Glowworm Swarm Optimization (GSO) Algorithm

GSO starts by distributing a swarm of agents randomly in the search space. Agents are modeled after glowworms and, hereafter, they will be called glowworms. Further, they are endowed with other behavioral mechanisms that are not found in their natural counterparts. Accordingly, the basic working of the algorithm is the result of an interplay between the following three mechanisms:

1. **Fitness broadcast**: Glowworms carry a luminescent pigment called *luciferin*, whose quantity encodes the fitness of their locations in the objective space. This allows them to glow at an intensity that is proportional to the function value being

© Springer International Publishing AG 2017
K.N. Kaipa and D. Ghose, *Glowworm Swarm Optimization*,
Studies in Computational Intelligence 698, DOI 10.1007/978-3-319-51595-3_2

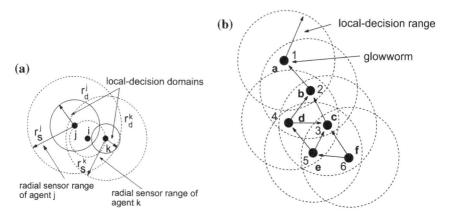

Fig. 2.1 **a** $r_d^k < d(i, k) = d(i, j) < r_d^j < r_s^k < r_s^j$. Agent i is in the sensor range of (and is equidistant to) both j and k. But, they have different decision-domains. Hence, only j uses the information of i. **b** Emergence of a directed graph based on the relative luciferin level of each agent and availability of only local information. Agents are ranked according to the increasing order of their luciferin values. For instance, the agent **a** whose luciferin value is highest is ranked '1' in the figure

    optimized. It is assumed that the luciferin level of a glowworm as sensed by its neighbor does not reduce due to distance.[1]

2. **Positive taxis**: Each glowworm is attracted by, and moves toward, a single neighbor whose glow is brighter than that of itself; when surrounded by multiple such neighbors, it uses a probabilistic mechanism (described in Sect. 2.1.1) to select one of them.

3. **Adaptive neighborhood**: Each glowworm uses an adaptive neighborhood to identify neighbors; it is defined by a local-decision domain that has a variable range r_d^i bounded by a hard-limited sensor range r_s $(0 < r_d^i \leq r_s)$. A suitable heuristic is used to modulate r_d^i (described in Sect. 2.1.1). A glowworm i considers another glowworm j as its neighbor if j is within the neighborhood range of i and the luciferin level of j is higher than that of i.

Note that the glowworms depend only on information available in the local-decision domain to decide their movements. For instance, in Fig. 2.1a, glowworm i is in the sensor range of (and is equidistant to) both j and k. However, j and k have different neighborhood sizes, and only j uses the information of i. Figure 2.1b shows the formation of a directed graph based on the relative luciferin level of each agent and on the availability of only local information. Each glowworm selects, using a probabilistic mechanism, a neighbor that has a luciferin value higher than its own and moves toward it. These movements, that are based only on local information and selective neighbor interactions, enable the swarm of glowworms to partition

---

[1]In natural glowworms, the brightness of a glowworm's glow as perceived by its neighbor reduces with increase in the distance between the two glowworms.

into disjoint subgroups that steer toward, and meet at, multiple optima of a given multimodal function.

The significant difference between GSO and most earlier approaches to multimodal function optimization problems is the adaptive local-decision domain, which is used effectively to locate multiple peaks.

2.1.1 Algorithm Description

The exposition of the algorithm is presented for maximization problems. However, the algorithm can be easily modified and used to find multiple minima of multimodal functions. GSO starts by placing a population of n glowworms randomly in the search space so that they are well dispersed. Initially, all the glowworms contain an equal quantity of luciferin ℓ_0. Each cycle of the algorithm consists of a luciferin update phase, a movement phase, and a neighborhood range update phase (Fig. 2.2). The GSO algorithm is given in Fig. 2.3.

Luciferin update phase: The luciferin update depends on the function value at the glowworm position. During the luciferin-update phase, each glowworm adds, to its previous luciferin level, a luciferin quantity proportional to the fitness of its current location in the objective function space. Also, a fraction of the luciferin value is subtracted to simulate the decay in luciferin with time. The luciferin update rule is given by:

$$\ell_i(t+1) = (1 - \rho)\ell_i(t) + \gamma J(x_i(t+1)) \tag{2.1}$$

where, $\ell_i(t)$ represents the luciferin level associated with glowworm i at time t, ρ is the luciferin decay constant ($0 < \rho < 1$), γ is the luciferin enhancement constant

Fig. 2.2 The phases that constitute each cycle of the GSO algorithm

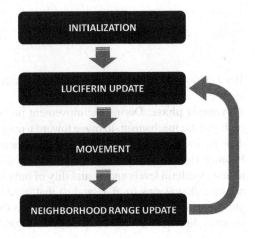

GLOWWORM SWARM OPTIMIZATION (GSO) ALGORITHM

Set number of dimensions $= m$
Set number of glowworms $= n$
Let s be the step size
Let $x_i(t)$ be the location of glowworm i at time t
$deploy\_agents\_randomly$;
for $i = 1$ to n do $\ell_i(0) = \ell_0$
$r_d^i(0) = r_0$
set maximum iteration number $= iter\_max$;
set $t = 1$;
while $(t \leq iter\_max)$ do:
{
    for each glowworm i do: % Luciferin-update phase
        $\ell_i(t) = (1 - \rho)\ell_i(t - 1) + \gamma J(x_i(t))$; % See Eq. (2.1)

    for each glowworm i do: % Movement-phase
    {
        $N_i(t) = \{j : d_{ij}(t) < r_d^i(t); \ell_i(t) < \ell_j(t)\}$;
        for each glowworm $j \in N_i(t)$ do:
            $p_{ij}(t) = \frac{\ell_j(t) - \ell_i(t)}{\sum_{k \in N_i(t)} \ell_k(t) - \ell_i(t)}$; % See Eq. (2.2)
        $j = select\_glowworm(\vec{p})$;
        $x_i(t + 1) = x_i(t) + s \left(\frac{x_j(t) - x_i(t)}{\|x_j(t) - x_i(t)\|} \right)$ % See Eq. (2.3)
        $r_d^i(t + 1) = \min\{r_s, \max\{0, r_d^i(t) + \beta(n_t - |N_i(t)|)\}\}$; % See Eq.
(2.4)
    }
    $t \leftarrow t + 1$;
}

Fig. 2.3 The GSO algorithm

and $J(x_i(t))$ represents the value of the objective function at glowworm i's location at time t.

Movement phase: During the movement phase, each glowworm decides, using a probabilistic mechanism, to move toward a neighbor that has a luciferin value higher than its own. That is, glowworms are attracted to neighbors that glow brighter. Figure 2.1b shows the directed graph among a set of six glowworms based on their relative luciferin levels and availability of only local information. For instance, there are four glowworms (a, b, c, and d) that have relatively higher luciferin level than glowworm e. Since e is located in the sensor-overlap region of c and d, it has only two possible directions of movement. For each glowworm i, the probability of moving toward a neighbor j is given by:

$$p_{ij}(t) = \frac{\ell_j(t) - \ell_i(t)}{\sum_{k \in N_i(t)} \ell_k(t) - \ell_i(t)} \tag{2.2}$$

where, $j \in N_i(t)$, $N_i(t) = \{j : d_{ij}(t) < r_d^i(t); \ell_i(t) < \ell_j(t)\}$ is the set of neighbors of glowworm i at time t, $d_{ij}(t)$ represents the Euclidean distance between glowworms i and j at time t, and $r_d^i(t)$ represents the variable neighborhood range associated with glowworm i at time t. Let glowworm i select a glowworm $j \in N_i(t)$ with $p_{ij}(t)$ given by (2.2). Then, the discrete-time model of the glowworm movements can be stated as:

$$x_i(t+1) = x_i(t) + s \left(\frac{x_j(t) - x_i(t)}{\|x_j(t) - x_i(t)\|} \right) \tag{2.3}$$

where, $x_i(t) \in R^m$ is the location of glowworm i, at time t, in the m-dimensional real space R^m, $\| \cdot \|$ represents the Euclidean norm operator, and s (>0) is the step size.

Neighborhood range update phase: Each agent i is associated with a neighborhood whose radial range r_d^i is dynamic in nature ($0 < r_d^i \le r_s$). The fact that a fixed neighborhood range is not used needs some justification. When the glowworms depend only on local information to decide their movements, it is expected that the number of peaks captured would be a function of the radial sensor range. In fact, if the sensor range of each agent covers the entire search space, all the agents move to the global optimum and the local optima are ignored. Since we assume that a priori information about the objective function (e.g., number of peaks and inter-peak distances) is not available, it is difficult to fix the neighborhood range at a value that works well for different function landscapes. For instance, a chosen neighborhood range r_d would work relatively better on objective functions where the minimum inter-peak distance is more than r_d rather than on those where it is less than r_d. Therefore, GSO uses an adaptive neighborhood range in order to detect the presence of multiple peaks in a multimodal function landscape.

Let r_0 be the initial neighborhood range of each glowworm (that is, $r_d^i(0) = r_0 \forall i$). To adaptively update the neighborhood range of each glowworm, the following rule is applied:

$$r_d^i(t+1) = \min\{r_s, \max\{0, r_d^i(t) + \beta(n_t - |N_i(t)|)\}\} \tag{2.4}$$

where, β is a constant parameter and n_t is a parameter used to control the number of neighbors.

The quantities ρ, γ, s, β, n_t, and ℓ_0 are algorithm parameters for which appropriate values have been determined based on extensive numerical experiments and are kept fixed in this book (Table 2.1). The quantity r_0 is made equal to r_s in all the experiments. Thus, n and r_s are the only parameters that influence the algorithm behavior (in terms of the total number of peaks captured) and need to be selected.

Table 2.1 Values of algorithmic parameters that are kept fixed for all the experiments

ρ	γ	β	n_t	s	ℓ_0
0.4	0.6	0.08	5	0.03	5

The following example is used to demonstrate the variable nature of the neighborhood. For the purpose of simplicity, a static placement of glowworms is considered and it is observed how the neighborhood range of the glowworm at $(0, 0)$ varies according to (2.4). In Figs. 2.4 and 2.5, notice that the neighborhood range gets adjusted until the glowworm acquires the number of neighbors that is specified by the parameter n_t $(=3)$. Figure 2.4a shows an initial placement where the glowworm at $(0, 0)$ is isolated. It increases its neighborhood range (Fig. 2.4b) until it either acquires a set of three neighbors or reaches the maximum range. In Fig. 2.5a, the glowworm is crowded by a large number of neighbors $(|N_i(t)| > n_t)$ that causes the neighborhood range to shrink until $|N_i(t)| = n_t$. Figure 2.5b shows the variation of the neighborhood range as a function of the iteration number.

When a glowworm is located far away from a peak, typically it has only a few neighbors or it may even be isolated (as in Fig. 2.4a). Rule (2.4) aids a glowworm in this situation to find neighbors by increasing its neighborhood range at each iteration. In contrast, when the glowworm is located in the vicinity of multiple peaks, it is more likely that the glowworm is surrounded by a large number of neighbors (as in Fig. 2.5a). Consequently, rule (2.4) restricts its visibility to a few neighbors so that its movement is biased toward a nearby peak.

2.1.2 Evolving Graph Architecture of GSO

The constraints of sensor range r_s and adaptive decision-range r_d^i, that are imposed on each glowworm, give rise to two different graphs[2]—an undirected graph N_c and a directed graph N_d—of the same set of glowworms. From the algorithm's description, it is clear that agents in GSO do not maintain fixed-neighbors during their movements, which means that new links may form, and existing links may break, between any two glowworms. Therefore, the graphs N_c and N_d are dynamic in nature. The example in Fig. 2.6 is used to describe the evolving graph architecture of agents in a glowworm swarm.

A group of 16 glowworms are randomly deployed in a search space that consists of three sources placed at different locations. Note from Fig. 2.6 that the graph N_c is connected. If the glowworms use a constant local-decision domain whose range is

---

[2]Let $G(V, E)$ be a graph with vertex set $V = \{v_1, \ldots, v_n\}$ and edge set $E = \{(v_i, v_j) : v_i, v_j \in V\}$. If E is a set of unordered pairs, then G is said to be an undirected graph. If E is a set of ordered pairs, then G is said to be a directed graph. The graph G is said to be connected if it has a path between each distinct pair of vertices v_i and v_j where by a path (of length m) is meant a sequence of distinct edges of G of the form $(v_i, k_1), (k_1, k_2), \ldots (k_m, v_j)$.

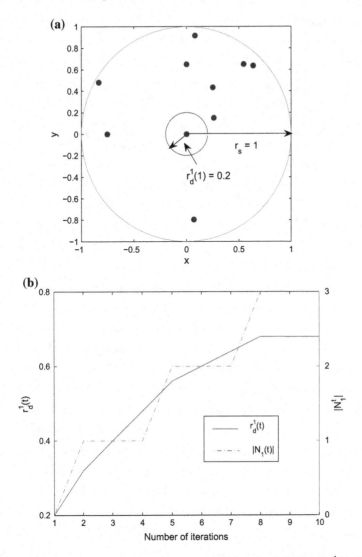

Fig. 2.4 **a** Initial placement where the glowworm at (0, 0) is isolated. **b** Plot of $r_d^1(t)$. Values of $r_s = 1$ and $n_t = 3$ are used

equal to the maximum sensing range r_s, all the glowworms converge to the global peak. This observation is supported by the simulation example presented in Sect. 2.4. However, note that the graph N_d is partitioned into two disjoint weakly connected components N_d^1 and N_d^2, which can be explained in the following way. When the glowworms use an adaptive decision-domain whose range is updated according to (2.4), the glowworms adjust their neighborhood ranges until they acquire a pre-specified number of neighbors ($n_t = 2$ in the example shown in Fig. 2.6). This property enables

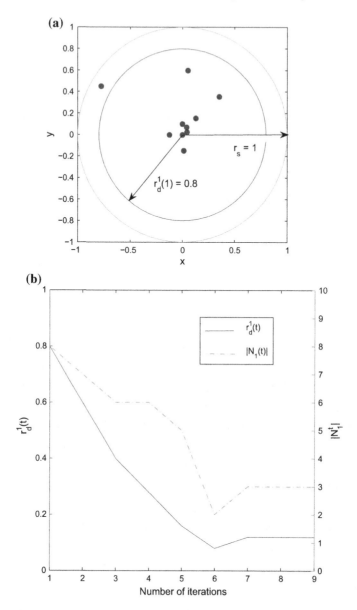

Fig. 2.5 **a** Initial placement where the glowworm at $(0, 0)$ is crowded by a large number of glow-worms. **b** Plot of $r_d^1(t)$. Values of $r_s = 1$ and $n_t = 3$ are used

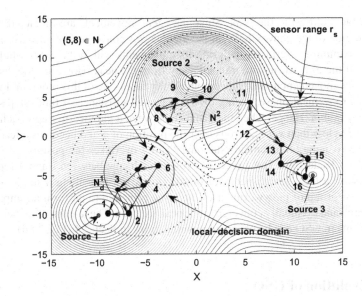

Fig. 2.6 Networks resulting from the constraints of sensing range r_s and adaptive decision-range r_d^i that are imposed on the agents in GSO

each glowworm to select its neighbors so that its movements get biased toward the nearest peak. This individual agent behavior results in a collective behavior of agents comprising an automatic splitting of the whole group into disjoint subgroups and the eventual convergence of each subgroup of agents to a nearby peak. In Fig. 2.6, glowworms 5, 6, 8, 9, and 10 are within the sensing-range of glowworm 7. However, glowworm 7 considers only glowworms 8 and 9 as neighbors. Therefore, its movements get biased toward Source 2, while it avoids moving toward glowworm 5, even though glowworm 5 has a higher luciferin value than that of itself. Accordingly, the subgroups of glowworms $\{1, 2, 3, 4, 5, 6\}$, $\{7, 8, 9, 10\}$, and $\{13, 14, 15, 16\}$ move toward Source 1, Source 2, and Source 3, respectively. Note that glowworms 11 and 12 may move toward either Source 2 or Source 3. The local-search and convergence of each subgroup to a nearby source location is achieved by the *leapfrogging* effect inherent in GSO, which is explained next.

2.1.3 Leapfrogging Behavior Enables Local Search

According to the basic GSO algorithm, in any given iteration the glowworm with the maximum luciferin remains stationary. The above property may lead to a dead-lock situation where all the glowworms in the vicinity of a peak converge to the glowworm that is located closest to the peak. Since the agent movements are restricted to the interior region of the convex-hull, all the glowworms converge to a glowworm that attains maximum luciferin value during its movements within the convex-hull. As a result, all the glowworms are trapped away from the peak. However, the discrete nature of the movement update rule automatically takes care of this problem. In fact, during the movement phase, each glowworm moves a distance of finite step size s toward a neighbor. Hence, when the distance between a glowworm i approaching a neighbor j is less than s, i leapfrogs over the position of j and becomes a leader to j. In the next iteration, i remains stationary and j overtakes the position of i thus regaining its leadership. This process of role interchange between i and j repeats, giving rise to a local-search behavior of the glowworm pair along a single ascent direction. A group of glowworms uses the same principle to perform an improved local-search and eventually converge to the peak location. The leapfrogging behavior of the agents in GSO can be observed in simulations in Sect. 2.4 (Fig. 2.8d).

2.2 Evolution of GSO

GSO, in its present form, has evolved out of several significant modifications incorporated into the earlier versions of the algorithm [103, 105, 107, 109, 112] that led to improvement in algorithmic performance from one version to the next. Some of the important steps in this evolution are briefly discussed for the sake of completion and also to convey the fact that many ideas were considered in the process of developing the GSO algorithm before converging upon the current version.

The algorithm was first introduced in [105]. The equations that modelled the luciferin-update, probability distribution used to select a neighbor, movement update, and local-decision range update are given below:

$$\ell_i(t+1) = \max\{0, (1 - \rho)\ell_i(t) + \gamma J(x_i(t+1))\} \tag{2.5}$$

$$p_j(t) = \frac{\ell_j(t)}{\sum_{k \in N_i(t)} \ell_k(t)} \tag{2.6}$$

$$x_i(t+1) = x_i(t) + s\left(\frac{x_j(t) - x_i(t)}{\|x_j(t) - x_i(t)\|}\right) \tag{2.7}$$

$$r_d^i(t+1) = \frac{r_s}{1 + \beta D_i(t)} \tag{2.8}$$

where,

$$D_i(t) = \frac{N_i(t)}{\pi r_s^2} \tag{2.9}$$

is the neighbor-density of agent i at iteration t and β is a constant parameter.

The local-decision range update rule (2.8) faces a problem at an instant when a glowworm i has no neighbors in its current local-decision domain range $r_d^i(t)$ but has some neighbors within its sensor range r_s. In particular, when the glowworm i is isolated, $D_i(t) = 0$. From (2.8), $r_d^i(t+1) = r_s$. Suppose glowworm i acquires some neighbors at $t+1$, $D_i(t+1) \neq 0$. If a large value of β is used, (2.8) gives $r_d^i(t+2) \approx 0$. Therefore, the above kind of neighbor-distribution associated with i results in an oscillatory behavior of the decision range, with r_d^i switching between r_s and a value closer to zero. Therefore, all glowworms in a neighborhood situation as described above move only in alternative iterations slowing down the convergence of the algorithm approximately by a factor of two.

To solve the above problem, the local-decision update rule in [106] is modified by forcing a nonzero lower bound on the decision range such that each glowworm improves its own chances of finding a neighbor at every instant. The modified update rule for $r_d^i(t)$ is given by:

$$r_d^i(t+1) = \alpha + \frac{r_s - \alpha}{1 + \beta N_i(t)} \tag{2.10}$$

where, α represents the lower bound of the decision domain range.

A new update rule is proposed in [104] where an explicit threshold parameter n_t is used to control the number of neighbors at each iteration. A substantial enhancement in performance is noticed by using this rule:

$$r_d^i(t+1) = \begin{cases} r_d^i(t) + \beta_1 |N_i(t)|, & \text{if } |N_i(t)| \leq n_t \\ r_d^i(t) - \beta_2 |N_i(t)|, & \text{otherwise} \end{cases} \tag{2.11}$$

where, β_1 and β_2 are constant parameters.

A further improvement in algorithmic performance is observed by using the present decision-range update rule (2.4), which was introduced in [107]. This can be attributed to the fact that the second term in (2.4), which either increments or decrements $r_d^i(t)$, is proportional to the difference between the desired number of neighbors n_t and the actual number of neighbors $|N_i(t)|$.

Since actual values of luciferin (instead of differences in luciferin values as used in the current version) are used in the probability distribution formula (2.6), the luciferin cannot take negative values. This is taken care of in the luciferin update formula (2.5), by forcing a zero lower bound on the luciferin value. However, the algorithm does not work in regions where the objective function has negative values unless the function is shifted appropriately. In order to address these problems, actual luciferin values in the probability distribution formula are replaced by relative luciferin values in [112]. As a consequence, the luciferin values are allowed to take negative values. Therefore, the luciferin-update formula was accordingly modified to its current form given in (2.1) where the constraint of zero lower bound on the luciferin values is removed [112].

2.3 Convergence Results

Some theoretical proofs on the working of the luciferin update rule are derived in this section. These theoretical results give an insight into how the luciferin levels of glowworms vary as a function of time during the execution of the algorithm. First, it is proved that, due to luciferin decay, the maximum luciferin level τ_{max} is bounded asymptotically. Secondly, it is shown that the luciferin ℓ_j of all glowworms co-located at a peak X_i converge to the same value ℓ_i^*.

Theorem 2.1 *Assuming that the luciferin update rule in (2.1) is used, the luciferin level $\ell_i(t)$ for any Glowworm i is bounded above asymptotically as follows:*

$$\lim_{t \to \infty} \ell_i(t) \leq \lim_{t \to \infty} \ell^{max}(t) = \left(\frac{\gamma}{\rho} \right) J_{max} \tag{2.12}$$

where, J_{max} is the global maximum value of the objective function.

Proof Given that the maximum value of the objective function is J_{max} and the luciferin update rule in (2.1) is used, the maximum possible quantity of luciferin added to the previous luciferin level at any iteration t is γJ_{max}. Therefore, at Iteration 1, the maximum luciferin of any Glowworm i is $(1 - \rho)\ell_0 + \gamma J_{max}$. At Iteration 2, it is $(1 - \rho)^2 \ell_0 + [1 + (1 - \rho)]\gamma J_{max}$, and so on. Generalizing the process, at any iteration t, the maximum luciferin $\ell^{max}(t)$ of any Glowworm i is then given by:

$$\ell^{max}(t) = (1 - \rho)^t \ell_0 + \sum_{k=0}^{t-1} (1 - \rho)^k \gamma J_{max} \tag{2.13}$$

Clearly,

$$\ell_i(t) \leq \ell^{max}(t) \tag{2.14}$$

Since $0 < \rho < 1$, from (2.13) we have that

$$\text{as } t \to \infty, \ell^{max}(t) \to \left(\frac{\gamma}{\rho} \right) J_{max} \tag{2.15}$$

Using (2.14) and (2.15), we have the result in (2.12). $\square$

Theorem 2.2 *For all glowworms i co-located at peak-locations X_j^* associated with objective function values $J_j^* \leq J_{max}$ (where, $j = 1, 2, \ldots, m$, with m as the number of peaks), if the luciferin update rule in (2.1) is used, then $\ell_i(t)$ increases or decreases monotonically and asymptotically converges to $\ell_j^* = \left(\frac{\gamma}{\rho} \right) J_j^*$.*

Proof According to (2.1), $\ell_i(t) \geq 0$ always. The stationary luciferin ℓ_j^* associated with peak j satisfies the following condition:

$$\ell_j^* = (1 - \rho)\ell_j^* + \gamma J_j^* \Rightarrow \ell_j^* = \left(\frac{\gamma}{\rho}\right) J_j^* \tag{2.16}$$

If $\ell_i(t) < \ell_j^*$ for Glowworm i co-located at peak-location X_j^*, then using (2.16) we have

$$J_j^*(t) > \left(\frac{\rho}{\gamma}\right) \ell_i(t) \tag{2.17}$$

Now,

$$\ell_i(t + 1) = (1 - \rho)\ell_i(t) + \gamma J_j^* \tag{2.18}$$

$$> (1 - \rho)\ell_i(t) + \gamma \left(\frac{\rho}{\gamma}\right) \ell_i(t)$$

$$\Rightarrow \ell_i(t + 1) > \ell_i(t) \tag{2.19}$$

that is, $\ell_i(t)$ increases monotonically.

Similarly, if $\ell_i(t) > \ell_j^*$ for Glowworm i co-located at peak-location X_j^*, then using (2.16) and (2.18), it is easy to show that

$$\ell_i(t + 1) < \ell_i(t) \tag{2.20}$$

that is, $\ell_i(t)$ decreases monotonically.

Now, the convergence of the sequence $\ell_i(t)$ is proved by showing that the fixed point ℓ_j^* of the system in (2.18) is asymptotically stable. From (2.18), the following relation can be deduced:

$$\left|\ell_i(t) - \frac{\gamma}{\rho}J_j^*\right| = (1 - \rho)\left|\ell_i(t - 1) - \frac{\gamma}{\rho}J_j^*\right| \tag{2.21}$$

$$= (1 - \rho)^2 \left|\ell_i(t - 2) - \frac{\gamma}{\rho}J_j^*\right|$$

Proceeding in a similar way, we get

$$\left|\ell_i(t) - \ell_j^*\right| = (1 - \rho)^t \left|\ell_i(0) - \ell_j^*\right| \tag{2.22}$$

Therefore,

$$\lim_{t \to \infty} \left|\ell_i(t) - \ell_j^*\right| = \lim_{t \to \infty} (1 - \rho)^t \left|\ell_i(0) - \ell_j^*\right|$$

$$= 0, \text{ since } 0 < (1 - \rho) < 1 \tag{2.23}$$

From (2.23), it is clear that the luciferin $\ell_i(t)$ of Glowworm i, co-located at a peak-location X_j^*, asymptotically converges to ℓ_j^*. $\qquad\square$

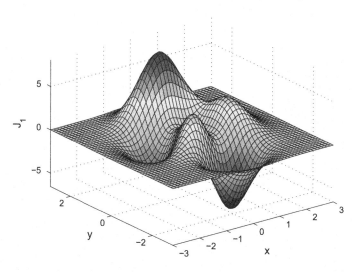

Fig. 2.7 Peaks function $J_1(x, y)$ [184] used in simulations to demonstrate the basic working of GSO. The function has three local maxima located at $(0, 1.58)$, $(-0.46, -0.63)$, and $(1.28, 0)$

2.4 Simulation Experiments to Illustrate GSO

Simulation experiments[3] demonstrating the capability of the glowworm algorithm to capture multiple peaks of a number of benchmark multimodal functions are deferred to Chap. 4. Here, the basic working of the algorithm is demonstrated using the *peaks* function, which is a function of two variables, obtained by translating and scaling Gaussian distributions [184]:

$$J_1(x, y) = 3(1 - x)^2 e^{-[x^2+(y+1)^2]} - 10\left(\frac{x}{5} - x^3 - y^5\right) e^{-(x^2+y^2)}$$
$$- \left(\frac{1}{3}\right) e^{-[(x+1)^2+y^2]} \tag{2.24}$$

The peaks function $J_1(x, y)$ has multiple peaks and valleys (Fig. 2.7). Local maxima are located at $(0, 1.58)$, $(-0.46, -0.63)$, and $(1.28, 0)$ with different peak function values. A set of 50 glowworms are randomly deployed in a two-dimensional workspace of size 6×6 square units.

---

[3]Movies of some of the simulations presented in this book can be viewed at this link: https://www.youtube.com/watch?v=_vhSu4xBoFs.

2.4.1 Simulation Experiment 1: Constant Local-Decision Domain Range

As a first step, the radial range r_d^i of each glowworm is kept constant, in order to characterize the sensitivity of the number of peaks detected to the size of the local-decision domain. As noted earlier, the local-decision range greatly influences the determination of various peaks. When the decision-range is more than 2, all the glowworms move to the global maximum. Figure 2.8a–c show the emergence of the solution, after 200 iterations, when the local-decision range r_d^i of all glowworms is kept constant at 2 (only one peak is captured), 1.8 (two peaks are captured), and 1.5 (all three peaks are captured), respectively.

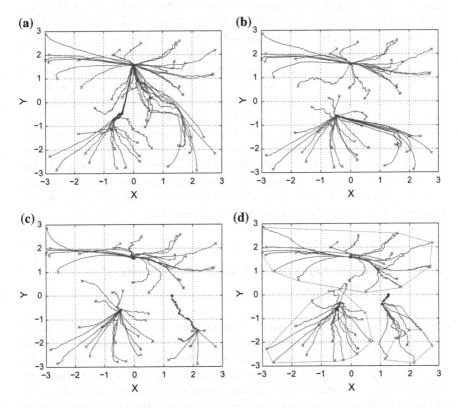

Fig. 2.8 Emergence of solution after 200 iterations for different cases: **a** The decision range is kept constant with $r_d^i = 2$ (only one peak is captured). **b** The decision range is kept constant with $r_d^i = 1.8$ (only two peaks are captured). **c** The decision range is kept constant with $r_d^i = 1.5$ (all three peaks are captured). **d** Decision range is made adaptive according to (2.4), with $r_d^i(0) = 3$ (all three peaks are captured)

2.4.2 Simulation Experiment 2: Adaptive Local-Decision Domain Range

Figure 2.8d shows the emergence of the solution when the local-decision domain range is made to vary according to (2.4) at each iteration t. During this simulation, a value of $r_d^i(0) = 3$ is chosen for each glowworm i. Note that all the peaks are detected within 200 iterations. In particular, 23, 19, and 8 glowworms get co-located at the maxima of (0, 1.58), (−0.46, −0.63), and (1.28, 0), respectively. Figure 2.9a shows the luciferin history of each glowworm. Note that after the steady state is reached, all the glowworms, co-located at a particular location, possess the same luciferin quantity. According to Theorem 2, the value of ℓ_j of all glowworms j, co-located at a peak-location X_i^*, is given by $(\frac{\gamma}{\rho})J_i^*$. It was observed that the value of ℓ at the three peak-locations (0, 1.58), (−0.46, −0.63), and (1.28, 0) are 12.15 (=(0.6/0.4) × 8.1), 5.67 (=(0.6/0.4) × 3.78), and 5.4 (=(0.6/0.4) × 3.6), respectively. Note that the luciferin values obtained in simulations are in exact agreement with their analytical values given by Theorem 2 (Fig. 2.9a).

Figure 2.9b shows the initial placement and co-location of all the glowworms on the equi-contour plot of the objective function. Figure 2.10a, b show the number of neighbors and the local-decision range of glowworm 12, respectively. Initially, the value of $r_d^{12}(t)$ decreases as glowworm 12 has more than five neighbors ($n_t = 5$). Note that between $t = 74$ and $t = 94$, the decision-range rises sharply up to the maximum sensing range r_s. The reason is evident from Table 2.2, where the number of glowworms that are within the decision-range of glowworm 12 is much higher than n_t, but among them, the number of glowworms that have a higher luciferin value (and hence, the number of neighbors) is less than n_t. However, at $t = 94$, the decision-range starts shrinking again, as it acquires a set of 13 neighbors.

2.4.3 Simulation Experiment 3: Effect of Presence of Forbidden Region

GSO has an advantage in situations where presence of forbidden regions in the workspace lead to loss of gradient information and local-gradient based algorithms fail to work. Figure 2.11 depicts a situation where the glowworm A is unaware of the source location because of the occlusion created by the forbidden region O. However, the inter-agent communication between neighbors makes global information available locally to the agent, providing the agent with feasible directions to move toward the source. This situation is simulated by biasing the random initial placement of the glowworms to ensure that none of the glowworms is deployed in a circular (forbidden) region of radius 1 unit centered at (0.6, −0.6). The size of the forbidden region

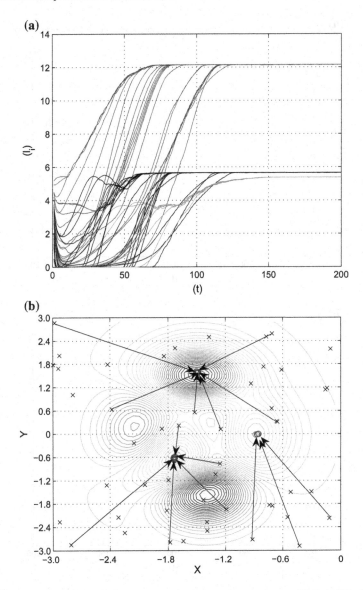

Fig. 2.9 a Luciferin-histories of the glowworms. **b** The equi-contour plot with the initial placement (shown by ×-*marks*) and final co-location (shown by the *three clusters of dots*) of the glowworms at the peak locations

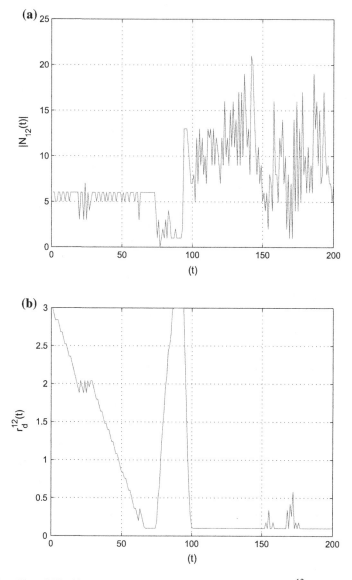

Fig. 2.10 **a** Plot of $|N_{12}(t)|$ as a function of iteration number t. **b** Plot of $r_d^{12}(t)$ as a function of iteration number t

Table 2.2 Positions and luciferin levels of glowworms that are within the local-decision range of glowworm 12 at $t = 74, 80$, and 94 iterations, respectively. The symbol $*$ on the superscript of a luciferin value of glowworm j represents that the corresponding glowworm is a neighbor of that glowworm 12 at that instant

Glowworm	$X_i(74)$	$\ell_i(74)$	$X_i(80)$	$\ell_i(80)$	$X_i(94)$	$\ell_i(94)$
1	$(-0.02, 1.60)$	12.059	$(-0.02, 1.55)$	12.144	$(0.01, 1.58)$	12.156*
3	$(-0.03, 1.60)$	12.136*	$(-0.02, 1.59)$	12.149	$(0.00, 1.58)$	12.155*
8	$(-0.04, 1.59)$	12.144*	$(-0.02, 1.58)$	12.144	$(0.00, 1.58)$	12.154*
12	$(-0.02, 1.59)$	**12.133**	$(-0.01, 1.56)$	**12.152**	$(-0.02, 1.57)$	**12.145**
16	$(0.15, 1.60)$	11.470	$(-0.01, 1.57)$	12.070	$(-0.02, 1.57)$	12.150*
19	$(-0.03, 1.60)$	12.142*	$(-0.02, 1.58)$	12.155*	$(-0.02, 1.58)$	12.158*
23	$(0.10, 1.61)$	11.716	$(-0.02, 1.58)$	12.119	$(0.00, 1.57)$	12.153*
24	$(0.13, 1.60)$	11.588	$(-0.03, 1.60)$	12.102	$(-0.02, 1.57)$	12.155*
26	$(0.29, 1.56)$	10.551	$(0.18, 1.58)$	11.575	$(0.00, 1.58)$	12.147*
31	$(-0.03, 1.60)$	12.131	$(-0.02, 1.59)$	12.148	$(-0.01, 1.57)$	12.151*
33	$(0.18, 0.16)$	11.361	$(0.00, 1.58)$	12.038	$(0.00, 1.57)$	12.151*
43	$(-0.04, 1.59)$	12.140*	$(-0.02, 1.59)$	12.146	$(0.00, 1.57)$	12.152*
44	$(0.037, 1.60)$	11.935	$(-0.01, 1.55)$	12.138	$(0.00, 1.58)$	12.153*
45	$(-0.03, 1.60)$	12.124	$(-0.02, 1.60)$	12.141	$(0.01, 1.59)$	12.151*

and sensor range are chosen such that a glowworm located on the boundary of the forbidden region cannot sense another glowworm that is located on the other side of the region. Also, the forbidden region is placed such that the peak located at $(1.28, 0)$ lies within the forbidden region. A constant local-decision range $(r_d^i(t) = 1.5, \forall t)$ that is equal to the maximum sensor range is used. The simulation result in Fig. 2.12a shows that, following the biased-random deployment as described above, there is no instance when a glowworm enters the forbidden region, and since one of the peaks is obscured by the forbidden region, only two peaks are detected. Figure 2.12b shows the glowworms that lie within the decision-domain of glowworm 26 at iteration $t = 1$. Since the region of intersection S (refer to Fig. 2.12b) of its decision-domain and the forbidden region is devoid of any neighbors, glowworm 26 avoids entrance into the forbidden region. Table 2.3 shows the initial positions and luciferin levels of glowworm 26 and its neighbors. Since it obtains feasible directions of movement at every time step, it continues to move by avoiding the forbidden region and finally gets co-located at the global-maximum.

Fig. 2.11 Situation where inter-agent communication helps a glowworm to select a feasible direction toward the source

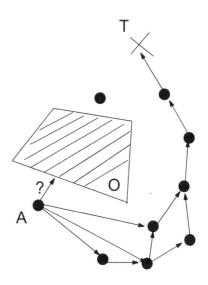

2.4.4 Effect of Step-Size on Convergence

In Fig. 2.12a, the peak-location $(-0.46, -0.63)$ lies outside the convex-hull C. Due to the constant step-size, all the glowworms climb the gradient-hill by performing the leapfrogging behavior as described in Sect. 2.1.3 and get co-located at the peak in their vicinity. Similar behavior of capturing a peak that lies outside the convex-hull of a group of glowworms can also be observed in the simulation result of Fig. 2.8d. In order to obtain the solution within the desired tolerance ϵ range, the step size s should be less than the tolerance value. However, this may lead to a slower convergence of the algorithm. This issue can be taken care of, by starting the algorithm with a coarse (large) step-size and progressively reducing the step-size, to a value lesser than ϵ, toward the convergence phase. For instance, when the step size $s(t)$ was updated as $s(t) = q^t s(0)$, with $q = 0.96$ and $s(0) = 0.2$, convergence was achieved within 50 iterations ($s(50) = 0.026$) as opposed to 200 iterations in the constant step-size case. However, this experimental result serves only as an illustrative example and various models for the adaptive step-size have to be explored in order to make it work on a wide variety of problems.

The step size is also a function of the size of the search space. For relatively large search spaces, we need to start with a larger step size. However, there is a chance of missing closely spaced peaks depending on the number of peaks in a given area. In these cases, an additional search procedure in each local region of a peak captured with a smaller step size could reveal the presence of more peaks.

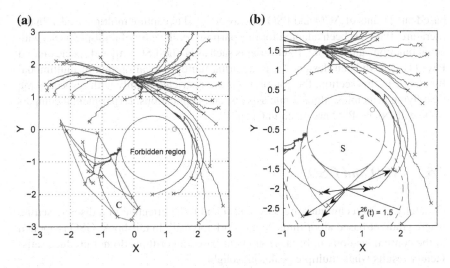

Fig. 2.12 **a** Response of the glowworm algorithm to the presence of a circular forbidden region of radius 1 unit centered at (0.6, −0.6). Emergence of the solution after 500 iterations; $r_d^i = 1.5$ **b** glowworm 26 has six feasible directions to move at Iteration 1

Table 2.3 Initial positions and luciferin levels of glowworm 26 and its neighbors in Fig. 2.12

Glowworm	$X_i(1)$	$\ell_i(1)$
6	(−0.5171, −2.6962)	2.7226
13	(−0.0405, −2.1075)	0.6371
21	(1,2484, −1.9892)	2.2977
26	**(0.5735, −2.0126)**	**0.6069**
32	(0.1413, −2.4149)	1.7583
36	(1.813, −1.5145)	2.9805
42	(−0.0651, −2.8011)	2.6762

2.5 Comparison of GSO with ACO and PSO

The GSO technique is a population-based algorithm and falls under the category of swarm intelligence methods. The algorithm shares some common features with the ant-colony optimization (ACO) and particle swarm optimization (PSO) algorithms but is different in many aspects that help in achieving simultaneous detection of multiple local optima of multimodal functions. This is a problem not directly addressed by ACO or PSO techniques. Generally, ACO and PSO techniques are used for locating global optimum. However, our objective is to locate as many of the peaks as possible. This requirement is the main motivation for formulating the GSO technique. Since the primary goal of GSO (that of capturing multiple peaks simultaneously) is different from that of ACO and PSO, comparisons with these techniques should be based on variants of ACO and PSO that are adapted to capture multiple peaks. In the literature, PSO has been adapted for this purpose for the case of multiple peaks with equal function values [168]. PSO variants such as Niche-PSO [20] and species-based PSO [117] have been designed for capturing multiple local optima of multimodal functions. In this section, the comparisons are restricted mainly with respect to the underlying principles, algorithmic aspects, and applications. Experimental comparisons with Niche-PSO are given in Chap. 4.

2.5.1 ACO and GSO

Generally, ACO techniques are used and found to be effective in a discrete setting where gradient-based algorithms do not work too well. In this book, GSO is applied to the continuous domain because gradient-based algorithms do not produce satisfactory results when multiple peaks are sought.

Traveling salesman problem (TSP) is a typical example of an ACO application where pheromone update on city routes closely mimics the trail-laying phenomenon found in foraging of ants. In general, most conventional ant-algorithms involve adding a pheromone value on specific routes where the agents visit. The pheromone level on these visited routes decay with time in order to emulate the evaporation behavior of actual pheromones. Even though the luciferin update mechanism draws inspiration from these stigmergic principles followed by ants, there is a significant modification in its implementation and how the luciferin information is used by the agents in GSO. The difference mainly arises from the fact that the luminescence does not stay at places visited by glowworms (unlike pheromones that remain associated with routes visited by ants), but moves along with them. Therefore, the luciferin level of a glowworm indicates the net improvement made by it while traversing a path that emerges from an initial location to its current location. A glowworm whose location makes relatively more net improvement in the objective function during its movements has a higher luciferin level and attracts more neighbors toward it than others, thus enabling agents to move toward favorable places of the environment. Clearly, the number of luminescence sources is equal to the number of glowworms. While the pheromonal decay in ACO algorithms serves to avoid premature convergence to, and eventual removal of, suboptimal paths/solutions, the luciferin decay component in GSO controls the luciferin values of the glowworms and ensures that the luciferin values are always bounded. Note that, when the solution is reached, the number of luminescence locations will be approximately equal to the number of maxima of the objective function as most of the agents settle at one of the multiple peaks.

Now, GSO is compared to a variant of ACO, which was introduced by Bilchev and Parmee [15] in order to address continuous optimization problems. In this ACO approach, a finite set of regions are randomly placed in the search space. Each path between the nest and a region i is associated with a virtual pheromone $\tau_i(t)$ at each iteration t. Initially, $\tau_i(t = 0) = \tau_0$ for all agents. The probability that an agent selects region i is given by:

$$p_i(t) = \frac{\tau_i^{\alpha}(t)\eta_i^{\beta}(t)}{\sum_{j=1}^{N} \tau_j^{\alpha}(t)\eta_j^{\beta}(t)} \tag{2.25}$$

where, $\eta_i(t)$ reflects the local-desirability of a portion of the solution, α and β represent relative weights, and N is the number of regions. The agent then moves to the selected region's center, measures the value of the objective function at that point, moves a short distance in a random direction, shifts the region's center to the new point if it finds an improvement in the solution, and then comes back to the nest. The pheromone update associated with the region is given by

$$\tau_i(t+1) = \begin{cases} (1-\rho)\tau_i(t) + \gamma \Delta J, & \text{if } \Delta J > 0 \\ (1-\rho)\tau_i(t), & \text{Otherwise} \end{cases} \tag{2.26}$$

where, ΔJ is the improvement made in the solution, ρ is the pheromone evaporation constant, and γ is a proportionality constant.

This process is repeated with a new probability distribution according to (2.25). With the increase in number of iterations, the pheromone concentration associated with *inferior* regions decay (and may disappear eventually) and *good* regions get reinforced with time, finally converging to the solution.

The virtual pheromones associated with the various regions in the above variant of ACO technique can be likened to the luciferin carried by the glowworms in the GSO technique. However, the crucial difference lies in the manner in which the stigmergic communication is used by the agents to make decisions. In this ACO variant, each agent at the nest selects a region based on a probability distribution (2.25) which is a function of the pheromone levels associated with all the N regions. In contrast, each glowworm in GSO broadcasts its own luciferin value to other agents and uses the luciferin information available only in its neighborhood to probabilistically select a neighbor with higher luciferin value. While the selected region's center is shifted, along a random direction, to a new point in this ACO variant, each glowworm deterministically moves one step toward the selected neighbor. The main differences between GSO and ACO are summarized in Table 2.4.

2.5.2 PSO and GSO

Particle swarm optimization (PSO) is a population-based stochastic search algorithm [29, 98]. In PSO, a population of solutions $\{X_i : X_i \in R^m, i = 1, \dots, N\}$ is evolved, which is modeled by a swarm of particles that start from random positions in the objective function space and move through it searching for optima; the velocity V_i of each particle i is dynamically adjusted according to the *best so far* positions visited

Table 2.4 Comparison of ACO and GSO

	Standard ACO	GSO
1	Effective in *discrete* setting [48]	Applied to *continuous* domain
2	Global optimum	Multiple optima of *equal* or *unequal* values
	Special variant of ACO [15]	
1	Cannot be applied when ants (agents) have limited sensing range	Useful for applications where robots have limited sensor range
2	*Global* information used	*Local* information used
3	Pheromones associated with *paths* from nest to regions	Luciferin carried by and associated with *glowworms*
4	Pheromone information used to select *regions*	Luciferin information used to select *neighbors*
5	Shifting of selected region's center in a *random* direction	*Deterministic* movements toward selected neighbor

by itself (P_i) and the whole swarm (P_g). The position and velocity updates of the ith particle are given by:

$$V_i(t + 1) = V_i(t) + c_1 r_1 (P_i(t) - X_i(t)) + c_2 r_2 (P_g(t) - X_i(t)) \qquad (2.27)$$
$$X_i(t + 1) = X_i(t) + V_i(t + 1) \qquad (2.28)$$

where, c_1 and c_2 are positive constants, referred to as the *cognitive* and *social* parameters, respectively, r_1 and r_2 are random numbers that are uniformly distributed within the interval $[0, 1]$, $t = 1, 2, \ldots$, indicates the iteration number. PSO uses a memory element in the velocity update mechanism of the particles. Differently, the notion of memory in GSO is incorporated into the incremental update of the luciferin values that reflect the cumulative goodness of the path followed by the glowworms.

GSO shares a feature with PSO: in both algorithms a group of particles/agents are initially deployed in the objective function space and each agent selects, and moves in, a direction based on respective position update formulae. While the directions of a particle movements in original PSO are adjusted according to its own and global best previous positions, movement directions are aligned along the line-of-sight between neighbors in GSO. In PSO, the net improvement in the objective function at the iteration t is stored in $P_g(t)$. However, in GSO, a glowworm with the highest luciferin value in a populated neighborhood indicates its potential proximity to an optimum location.

Figure 2.13a–c illustrate the basic concepts behind PSO and GSO and bring out the differences between them. Figure 2.13a shows the trajectories of six agents, their current positions, and the positions at which they encountered their personal best and the global best. Figure 2.13b shows the GSO decision for an agent which has four neighbors in its dynamic range of which two have higher luciferin value and thus only two probabilities are calculated. Figure 2.13c shows the PSO decision, which

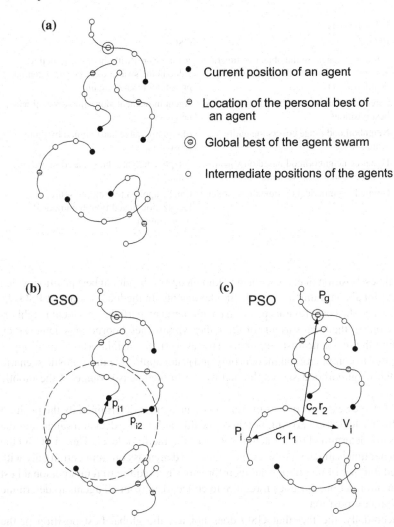

Fig. 2.13 **a** Trajectories of six agents, their current positions, and the positions at which they encountered their personal best and the global best. **b** GSO decision for an agent which has four neighbors in its dynamic range of which two have higher luciferin value and thus only two probabilities are calculated. **c** PSO decision which is a random combination between the current velocity of the agent, the vector to the personal best location, and the vector to the global best location [113]

is a random combination between the current velocity of the agent, the vector to the personal best location, and the vector to the global best location.

In PSO, the next velocity direction and magnitude is dependent on combination of the agent's own current velocity, and randomly weighted global best vector and personal best vector. While this is implementable in a purely computational platform, implementation of this algorithm in robotics platform or a platform containing realistic agents would demand large speed fluctuations, presence of memory of the

Table 2.5 PSO versus GSO

	PSO	GSO
1	Net improvement in the objective function at iteration t stored in the global best position $P_g(t)$	A glowworm with the highest luciferin value in the swarm indicates its potential proximity to an optimum
2	Direction of movement based on previous best positions	Agent movement along line-of-sight with a neighbor
3	Neighborhood range covers the entire search space	Maximum range hard limited by finite sensor range
4	Dynamic neighborhood based on k nearest neighbors (in a local variant of PSO)	Adaptive neighborhood based on varying range
5	Limited to numerical optimization models	Can be applied to multiple source localization in addition to numerical optimization

personal best position of each agent, and knowledge of the global best position which requires global communication with all other agents. In the local variant of PSO, P_g is replaced by the best previous position encountered by particles in a local neighborhood. In one of the local variants of PSO, the dynamic neighborhood is achieved by evaluating the first k nearest neighbors. However, such a neighborhood topology is also limited to computational models only and is not applicable in a realistic scenario where the neighborhood size is defined by the limited sensor range of the mobile agents.

In GSO, the next direction of movement of an agent is determined by the position of the higher luciferin carrying neighbors within a dynamic decision range and the weights are determined by the actual values of the luciferin level. Thus, in GSO the implementation is much simpler as the algorithm demands communication only with a limited number of neighbors and therefore does not require to retain personal best position information, nor does it require to collect data from all agents to determine the global best position.

Conceptually, the fact that GSO does not use the global best position or the personal best position and the fact that it uses information only from a dynamic neighbor set helps it to detect local maxima, whereas PSO gets easily attracted to the global maxima.

The above figures and explanation imply that GSO is completely different from PSO although they have some minor commonalities. The main differences between GSO and PSO are summarized in Table 2.5.

2.6 Summary

In this chapter, the development of glowworm swarm optimization (GSO) algorithm was presented. The underlying ideas behind the GSO technique, the notion of an adaptive local-decision domain, and the steps involved in the implementation of the basic GSO algorithm were described. Some theoretical results were presented that gave an insight into how the luciferin levels of glowworms vary as a function of time. Later, a simulation example was used to illustrate the basic capability of GSO for solving multimodal function optimization problems; several aspects of GSO like the effect of using an adaptive local-decision domain, local search due to leapfrogging behavior, and the effect of forbidden region on agent behavior were demonstrated. Finally, some comparisons of GSO with other bio-inspired optimization techniques were described. In the next chapter, the theoretical performance of a simplified GSO model and some illustrative simulations to support the theoretical findings are presented.

2.7 Thought and Computer Exercises

Exercise 2.1 The original GSO algorithm is described for solving maximization problems. Show that a minor modification to the definition of a neighbor is sufficient to extend GSO to minimization problems. Can you think of any other method to achieve the same purpose? (Hint: Consider changing the sign of the terms in the luciferin update rule given in (2.1)). Support your claims through numerical simulations (Use the MATLAB code provided in the Appendix). For this purpose, first run the 'GSO.m' file and verify that it captures the three maxima of the Peaks function (2.24), which is specified in the 'UpdateLuciferin.m' file. Now, update the code with each modification and confirm if the algorithm is able to capture the minima of this multimodal function. Comment on your observations.

Exercise 2.2 The parameter ρ (luciferin decay constant) in the luciferin update rule (2.1) simulates the decay of luciferin with time. What value of ρ results in a memoryless variant of GSO? Run the GSO code on a few multimodal test functions for different values of ρ while keeping other parameters constant. Analyze the impact of ρ on the emergence of solution in each case via different plots. Does the impact of ρ change if the parameter n (number of agents) is varied?

Exercise 2.3 In the movement update of GSO, each glowworm uses a probabilistic mechanism (2.2) to select a brighter neighbor and moves a small step closer to it (2.3). A new variant of GSO can be obtained by letting the glowworm to deterministically select the max-neighbor (refers to the glowworm with the maximum luciferin value in its neighborhood) instead of moving based on probability. Explain via different simulation scenarios how the emergence of solutions changes when this GSO variant is used.

Exercise 2.4 The parameter n_t sets the desired number of neighbors for each glowworm (2.4). Comment on the impact of n_t on the connectivity of the agent graph. For example, what happens when $n_t = 0$? What happens when $n_t = n - 1$? Use simulation examples to support your arguments.

Exercise 2.5 Consider the multimodal function given below (2.29) in a search space $[-3, 3] \times [-3, 3]$:

$$J(x, y) = 3 \left(e^{-b[(x+\frac{d}{2})^2 + y^2]} - e^{-b[x^2 + y^2]} + e^{-b[(x-\frac{d}{2})^2 + y^2]} \right) \qquad (2.29)$$

2.5.1 The above function has two peaks at $(-\frac{d}{2}, 0)$ and $(\frac{d}{2}, 0)$. The variable d controls the distance between the two peaks. The function profile for various values of d is shown in Fig. 2.14. Run the GSO code for $d = 1$ and $b = 10$. Use Table 2.1 to set the values of ρ, γ, β, n_t, s, and ℓ_0. Set $n = 100$ and $r_s = 3$. Terminate the run after 1000 iterations (This large value is used only to ensure proper convergence. A better set of terminal conditions will be described in Chap. 4). Is the algorithm able to capture both the peaks? If yes, compute the solution error w.r.t. each peak (Hint: Solution error may be given by the distance between the true peak location and the average location of all the glowworms co-located at that peak. A more formal definition is deferred to Chap. 4).

2.5.2 Decrease d in steps of 0.1 and repeat the run for each case. Plot the solution error for each peak as a function of d. Comment on your observations. What is the minimum value of d for which the algorithm is able to capture both the peaks within a error tolerance of 0.05?

2.5.3 Set $d = 0.1$ and repeat the run. What do you observe? Explain why this happens (Hint: Plot the multimodal function for this value of d).

2.5.4 The parameter b controls how the slope of the function profile changes in the vicinity of each peak. Set $b = 50$ and plot the multimodal function again. Notice the presence of two peaks now. Verify if the algorithm is able to capture the two peaks in this new regime. Comment on your findings. If the algorithm fails, can you suggest any changes to the GSO parameters that enable the algorithm to distinguish between the peaks? (Hint: Keep decreasing n_t in steps of 1 and see what happens).

Exercise 2.6 Consider a modification of the Two-peaks function used in the previous exercise as shown below:

$$
\begin{aligned}
J(x, y) = {} & 6e^{-10[(x-0.5)^2 + (y-0.5)^2]} - e^{-10[x^2 + y^2]} \\
& + 4e^{-10[(x-0.5)^2 + (y+0.5)^2]} + 3e^{-10[(x+0.5)^2 + (y+0.5)^2]} \\
& + 2e^{-10[(x+0.5)^2 + (y-0.5)^2]}
\end{aligned}
\qquad (2.30)
$$

By inspection of (2.30), it is clear that the above function has four peaks. Identify these peak locations. Where does the global peak occur? Now, run the GSO code

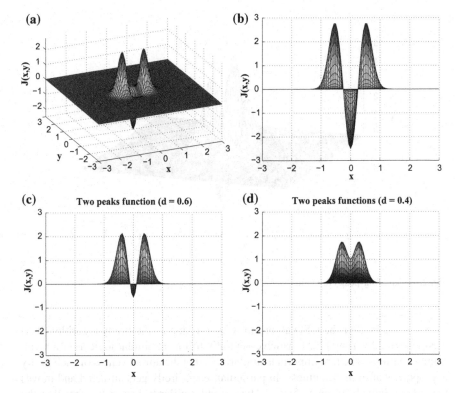

Fig. 2.14 The Two-peaks function for various values of d: **a** $d = 1$ (3D plot). **b** $d = 1$ (2D plot). **c** $d = 0.6$. **d** $d = 0.4$. The value of b is fixed at 10 in all three cases

using the nominal GSO parameters used in the previous exercise for the following two regimes:

1. Adaptive neighborhood case: Run the original GSO with $r_s = 3$ and record the number of peaks captured.
2. Constant neighborhood case: Modify (2.4) to $r_d^i(t + 1) = r_s$ (This change can be made in 'Act.m', line 17) and run the modified GSO for fixed values of r_s (=3, 2, 1, and 0.5). Record the number of peaks captured in each case.

Use your observations to comment on the impact of adaptive neighborhood on the performance of GSO (Fig. 2.15).

Additional Notes

The basic GSO algorithm has been modified by several researchers subsequent to its first appearance. Some of these works have suggested modifications in the basic algorithm and some have combined it with other swarm intelligence algorithms to get improved performance. A good summary of these modifications and hybridizations has been described by Singh and Deep [197]. Some of these will be discussed in Chap. 8 with more details.

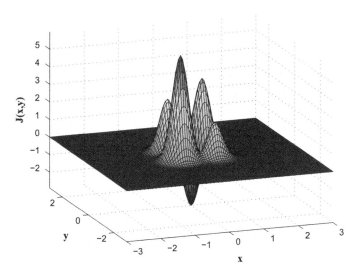

Fig. 2.15 The four-peaks function

A few years after the GSO algorithm was introduced, the firefly algorithm (FA) was proposed by Yang [222], which essentially follows a similar logic as GSO with some variations. The first important similarity is that the attractiveness of each firefly is proportional to its brightness. In particular, each firefly gets attracted and moves toward a relatively brighter firefly. The second similarity lies in the fact that the brightness of the glow of each firefly is determined by the landscape of the objective function. The FA algorithm has two variations. In GSO, the attractiveness of each glowworm is constant within a fixed sensing range r_s and zero beyond the range. However, the attractiveness of each agent in FA exponentially decays with distance. The second variation is the addition of a small randomization into the movement update of each firefly.

The key property (adaptive neighborhood) of GSO that enables an explicit splitting behavior of the swarm is not present in FA. The effect of the exponentially decaying property of agents' brightness in FA is that, although farther neighbors have less influence on the movement-decisions of each agent, it is still fully connected with all the remaining agents in the swarm, thereby leading to a centralized system. This makes FA more geared toward global optimization.

Work on GSO that appeared in the recent literature can be primarily classified into three categories. Some researchers proposed modifications of GSO [3, 4, 50, 75–77, 165, 171, 203, 219, 243, 244], most of which (with the exception of [3, 4]) were geared toward modifying GSO for global optimization problems. Some used basic GSO in different applications [27, 33, 90, 100, 135, 214], while others modified GSO and used them in some applications [3, 89, 116, 144, 147, 148, 213, 226, 235, 242].

Although GSO was designed for multimodal optimization, it was used for global optimization by Chetty and Adewumi [27]. The authors showed that its performance is comparable to other swarm intelligence algorithms that are specifically designed for global optimization. They used the effectiveness of employing these algorithms on an annual crop planning problem as a metric for comparison purposes.

Ouyang et al. [165] proposed BFGS-GSO, a hybrid algorithm of Broyden–Fletcher–Goldfarb–Shanno (BFGS) algorithm, which is one of the well-known classical gradient based algorithms, and GSO for global optimization problems. The authors showed that BFGS-GSO improved on the performance of basic GSO on a set of eight standard benchmark test functions.

Zhou et al. [243] proposed a GSO variant, in which the swarm was divided into two sub-populations that co-evolved toward global optima. Agents in one sub-population executed the basic GSO, while those in the other sub-population executed an algorithm based on Lévy flights. Lévy flights refer to a class of generalized random walks in which the step lengths during the walk are described by a 'heavy-tailed' probability distribution [12].

Du et al. [50] modified GSO for global optimization by omitting the adaptive neighborhood and probability based neighbor-selection phases and adding an element of elitism. The authors showed that the simplified-GSO performed better than the basic GSO and other swarm intelligence algorithms on a set of nine benchmark multimodal functions.

Zhouab et al. [244] incorporated principles of artificial fish swarm algorithm (AFSA) and differential evolution into GSO and applied it to constrained global optimization problems. A local search strategy based on simulated annealing was also applied in order to overcome premature convergence. The authors used tests on several benchmark functions to show that these modifications improved the convergence efficiency and computational precision of GSO.

He et al. [76] proposed a GSO variant by incorporating a two-layer hierarchical structure into the basic GSO algorithm. The bottom layer consisted of multiple subswarms of glowworms that separately searched different partitions of the search space. The optima captured in multiple partitions were used to initialize a new swarm of glowworms that formed the top layer of the hierarchy. The operators of selection and crossover were incorporated into the top-level swarm which led to enhancement in the diversity of the swarm. The authors reported simulation results on several benchmark functions to show that these modifications improved the speed and accuracy of GSO.

Aljarah and Ludwig [3, 5] parallelized GSO by using MapReduce, a popular programming model developed by Google for processing and generating large data sets with a parallel, distributed algorithm on a cluster [38]. The authors showed significant speedup and scalability, while maintaining the optimization quality on several large scale multimodal functions. The clustering-GSO was also parallelized by using MapReduce by Al-madi et al. [2] and similar improvements in speed and scalability were reported.

Huang and Zhou [84] introduced chaotic search strategies in GSO to obtain a more well-distributed initial solution. Zhou et al. [239] modified GSO by the cloud model of optimization to improve its convergence property. Gu and Wen [69] attempted to improve the convergence property of GSO by using quantum models which increases the diversity of the swarms. Singh and Deep [196] proposed several variants of the original GSO on the basis of step size variations and tested them against standard benchmark problems. Similarly, Zhang et al. [230] presented an adaptive step size glowworm swarm optimization algorithm, which is claimed to improve convergence of the original GSO.

Basic GSO was used by Manimaran and Selladurai [135] to solve nonlinear fixed charge transportation problem in a single stage supply chain network. The objective function consists of a fixed cost that is incurred for every route and a variable cost that is proportional to the amount shipped. Difficulty in solving the problem arises due to nonlinearities in variable costs and discontinuities caused due to fixed costs. GSO was formulated to compute a least cost transportation plan that minimizes the total variable and fixed costs while satisfying the supply and demand requirements of each plant and customer. The authors showed that GSO performs better than spanning tree-based genetic algorithm in terms of total distribution cost.

Couceiro et al. [33] conducted swarm robotic experiments to benchmark five state-of-the-art algorithms for cooperative exploration tasks: (1) Robotic Darwinian Particle Swarm Optimization (RDPSO) [31, 32], (2) Extended Particle Swarm Optimization (EPSO) [178, 179], (3) Physically-embedded Particle Swarm Optimization (PPSO) [78], (4) GSO, and (5) Aggregations of Foraging Swarm (AFS) [60, 61]. All the five algorithms were described as belonging to a class of algorithms originally designed to solve optimization problems and later adapted to embrace the principles associated with real robotic tasks. Features like obstacle avoidance, initial deployment, communication mechanism, parametrization, handling of multiple/dynamic sources, computational complexity, memory complexity, and communication complexity were used to evaluate the theoretical advantages and disadvantages of the algorithms. The authors noted that only RDPSO and GSO were suited to handle multiple and dynamic sources. The authors initially compared the algorithms in a multi-robot simulation where agents collectively explored a large basement garage with a large density of obstacles. Metrics like exploration ratio and the associated area under the curve were used to evaluate each algorithm. The effectiveness of the top three performers found in simulations (RDPSO, GSO, and AFS) was further explored in real experiments using a swarm of fourteen e-puck robots [154]. The task consisted of collectively finding two victims emulated by e-pucks located at diagonally opposite corners of the workspace. The time taken to rescue each victim was used as a metric to evaluate each algorithm. The authors showed that RDPSO performed best and that GSO's performance closely followed RDPSO. The authors argued that this was a crucial result as GSO presents itself as a "low cost" alternative to the RDPSO in terms of computational and memory requirements. They also noted that although RDPSO performed better than GSO, a similar final outcome would be achieved by GSO if a larger mission time were available.

Senanayake [191] compared various swarm intelligence algorithms, including GSO, for the problem of search and tracking by a swarm of robots. This was a fairly exhaustive comparison.

Zhou et al. [241] used a modified GSO with a random operator to solve the problem of scheduling the dispatch of public transport vehicles.

Jiang and Tan [90] applied basic GSO to optimization of polarimetric multiple-input multiple-output (MIMO) radar systems. The problem involves a distributed array of transmitters and receivers used to detect a target. The goal is to select the parameters of the polarization waveforms of the transmitter array that maximize the probability of detection of the target. The authors formulated the optimization problem in the framework of GSO by using the formula for probability of detection in the luciferin update function and allowing the glowworms to search in the space of polarization parameters. The authors showed that the proposed GSO based algorithm outperformed other transmit waveform polarization schemes.

Liang et al. [119] applied GSO to solve a combined system identification and adaptive control problem for a mechanical servo system.

Xing [220] explored the possibility of using various computational intelligence methods, including GSO, to various robot related activities, such as location identification, manipulation, communication, vision, learning, and docking capabilities, in the context of assisted living for elderly people. The same author also tested computational intelligence algorithms (including GSO) for their suitability for data mining problems in the context of assisted living [221].

Kavipriya [96] applied GSO to solve code allocation problems in communication channels.

Kang et al. [94] applied GSO to solve a problem in X-ray based navigation. The bispectral feature points of the standard pulsar integrated pulse profile is extracted by the GSO algorithm and stored in the spacecraft's database. This information is now used to compute the X-ray pulsar time delay. The method shows a very good improvement in terms of computation time and suppression of Gaussian noise.

Pushpalatha and Ananthanarayana [180] proposed a GSO based document clustering algorithm that helps to cluster documents according to topics. The algorithm was tested through cluster based retrieval of multimedia documents.

Reddy and Rathnam [183] addressed the problem of optimizing power flow by minimizing the generation cost while keeping the power outputs of generators, bus voltages, bus shunt reactors/capacitors and transformer tap settings within acceptable limits. The problem is formulated as a multi-objective optimization problem keeping the minimization of emission for environmental reasons in mind. Both PSO and GSO is used to obtain useful results. Mageshvaran and Jayabarathi [129] used GSO to minimize the amount of load shedding in power systems in order to contain the deleterious effect of cascaded tripping and blackout. Wang et al. [214] used GSO for optimizing load allocation between hydropower units. They formulated the problem as a multi-objective optimization model in accordance with the characteristics and particularity of each station, with the minimum water rate of the station as the optimal objective.

Khan and Sahai [100] applied GSO for adaptive usability evaluation of B2C eCommerce web services.

Tang and Zhou [202] modified GSO by combining it with some aspects of PSO and applied it to path planning of uninhabited combat air vehicles. They showed the effectiveness of GSO by comparing it with ten other population-based optimization methods.

Zhao et al. [236] incorporated the Dijkstra's shortest path algorithm and genetic operators from GA into GSO to solve shortest path problems. Don et al. [43] used an adaptive discrete GSO algorithm and compared its performance by extensive simulations on the Travelling Sales Problem (TSP) with ACO and PSO.

Aljarah and Ludwig [4] modified GSO for data clustering applications. Specifically, the fitness function used to evaluate the goodness of the glowworms was adjusted to locate multiple optimal centroids. The fitness function was set to maximize similarity of glowworms within a cluster (by minimizing intra-agent distance) and minimize similarity of glowworms lying in different clusters (by maximizing inter-agent distance). The authors used the entropy and purity metrics to evaluate algorithmic performance. On most of data sets (Iris, Ecoli, Glass, Balance, Seed, Mouse, and VaryDensity) selected from the UCI database repository,[4] the authors showed that their clustering-GSO outperformed other well-known clustering algorithms like K-Means clustering [128], average linkage agglomerative Hierarchical Clustering (HC) [237], Furthest First (FF) [81], and Learning Vector Quantization (LVQ) [102].

In the work carried out by Marinaki and Marinakis [137], GSO has been used in conjunction with other meta-heuristic algorithms to solve a vehicle routing problem with random demands. This is a case where GSO has been used to solve a combinatorial optimization problem. An innovative method, to express the solution as a combination of two vectors, one in the continuous solution space and the other in the discrete solution space, is proposed. The former is updated using GSO while the latter uses a combinatorial neighbourhood topology technique.

Yepes et al. [225] addressed the problem of optimizing cost and CO_2 emissions while designing pre-cast, pre-stressed, concrete road bridges. GSO has been used in a hybrid scheme with simulated annealing (SA), where the local searches are carried out by SA and the global one by GSO. The problem is defined in a 40 dimensional decision space.

Cui et al. [35] proposed a modification to GSO in terms of making the weight on the location update rule vary, followed by a hybrid scheme that uses differential evolution in conjunction with GSO for solving the problem of time series prediction using single multiplicative neuron to which feedback and feedforward links are incorporated. The author claims that this modification improves the performance of GSO by avoiding locally optimal traps and overcomes the deficiency of not using memory of the search history.

Li and Huang [118] solved the problem of blind signal separation using a modified version of GSO based on step-size adjustment. Pubo and Yu [177] also dealt with blind source separation problem using GSO and its modification by baffle effect.

Jayakumar and Venkatesh [89] proposed two modifications of the original GSO to solve a multi-objective optimization problem related to environmental economic dispatch. In particular, the multiple objectives included minimization of fuel costs and

---

[4]http://archive.ics.uci.edu/ml/index.html.

minimization of emissions, respectively. The first modification consisted of using TOPSIS (Technique for Order Preference Similar to an Ideal Solution), a multi-criterion decision making method to compute the fitness of each glowworm. For this purpose, each glowworm is evaluated with respect to all the objective functions and ranked using the TOPSIS method, which works by assigning a higher rank to agents that simultaneously have shorter geometric distance from the positive ideal solution and farther geometric distance from the negative ideal solution. The second modification was a time varying step-size as opposed to a fixed step-size in the original GSO algorithm. The authors showed that the proposed GSO-TOPSIS approach produced comparable results over other state-of-the-art algorithms, and better results in some cases.

Meher et al. [144] modified GSO by incorporating stochastic updating of glowworm positions and applied it to an optimization problem in power systems. The objective of the economic load consignment problem was to obtain the most favorable active power outputs that minimize the generating cost while gratifying the running constraints. The authors evaluated the algorithm on a set of fourteen power generation models and showed it provided better results than those using differential evolution (DE) found in the literature [133]. Later, the same authors solved the dynamic load dispatch problem of thermal generating units using GSO [145] where the objective was to schedule optimal power generation of dedicated thermal units over a specific time band.

Unit commitment (UC) is a problem in power systems whose objective is to determine the on/off status of each power unit and the economic dispatch of power demand in a scheduling period so as to minimize the total system production cost subject to constraints of the units and the power system. Li et al. [116] developed a binary version of GSO and applied it to the UC problem. Each glowworm represented a matrix of on/off status of all the units at every hour, over the entire scheduling period. The Euclidean distance metric in the original GSO was replaced by the Hamming distance. The algorithm was shown to be very competitive in solving the UC problem in comparison to previously reported results by algorithms like quantum-inspired evolutionary algorithm, improved binary particle swarm optimization, and mixed integer programming.

Zhou et al. [238] addressed the problem of optimal sensor placement for structural health monitoring systems. The problem is posed as a multi-objective optimization of information entropy indices, which provide uncertainty metrics for the identified structural parameters. The basic GSO with some modifications using binary coding system is proposed. The application area is structural health monitoring of a long-span suspension bridge.

Wang et al. [213] modified GSO by incorporating a congestion factor and used it to optimize the parameters of an echo-state network (ESN)[5] based soft-sensor model of flotation processes.

Yu and Yang [226] used GSO with an adaptive step-size for a scheduling problem. They considered the whole-set orders problem: different customers place orders,

---

[5]ESN is a recurrent neural network with sparsely connected neurons.

where each order consists of multiple workpieces with different processing times and a different overall completion deadline. The goal is to maximize the number of weighted wholeset-orders. The authors reported that GSO performed better than genetic algorithms on this problem.

Menon and Ghose [147] modified GSO to address the problem of localizing the sources of contaminants spread in an environment, and mapping the boundary of the affected region. The authors used two types of agents, called the source localization agents (or S-agents) and boundary mapping agents (or B-agents) for this purpose. They defined new behaviour patterns for the agents based on their terminal performance as well as interactions between them that help the swarm to split into subgroups easily and identify contaminant sources as well as spread along the boundary to map its full length. Later the authors extended this GSO model to boundary mapping of 3-dimensional regions [148].

Zhou et al. [242] proposed a K-means image clustering algorithm based on GSO and showed its effectiveness in performing image classification on several benchmark images. The authors discuss the drawbacks of classical K-means clustering (sensitivity to initial conditions and getting trapped in local optima) and how GSO can do well in global searching, by searching for optimal clusters in parallel, thereby avoiding the impact of initial conditions. The authors reported that GSO performed better than K-means and fuzzy C-means (FCM) on the chosen benchmark images.

Zhao et al. [235] used a modified version of GSO for optimizing parameters of a Kaplan turbine (a propeller-type turbine with adjustable blades) with the goal of decreasing hydraulic losses and maximizing operating efficiency. The authors used a combination of a Relevance Vector Machine model and GSO to approximate the relationship between the wicket gate opening and runner blade angle.

Raman and Subramani [181] considered the problem of prioritization and minimization of the number of test cases for software testing. The authors proposed a modified GSO to solve this problem by conceptualizing a definite updating search field in the movement rule of the original GSO. The objectives were to maximize the path coverage and fault coverage to obtain optimal prioritized test cases. The authors claimed that the resulting solution guaranteed an optimal ordering of test cases and compared the performance of their modified GSO with PSO and artificial bee colony optimization (BCO).

Mo et al. [152] presented an example from economics and finance where GSO is used to determine the parameters of the Black–Scholes option pricing model.

Chapter 3
Theoretical Foundations

The previous chapter described the development of GSO and its basic ability to optimize multimodal functions. This chapter deals with the theoretical foundations for GSO. Initially, a characterization of the swarming behavior of agents in GSO is carried out, which guides the formulation of a framework used to analyze the algorithm. A theoretical model of GSO is obtained by using assumptions that make it amenable to analysis, yet reflecting most of the features of the original algorithm. Next, local convergence results for this GSO model are provided: First, when the glowworms are under the influence of an *isolated leader*, an upper bound on the time taken by the glowworms to converge to the isolated leader is determined. Second, when the glowworms are under the influence of two leaders with *non-isolated* and *non-overlapping* neighborhoods, an upper bound on the time taken by the glowworms to converge to one of the leaders is determined. Third, it is shown that glowworms, under the influence of two leaders with overlapping neighborhoods, asymptotically converge to one of the leaders. Some illustrative simulations are presented to support these theoretical findings.

3.1 Characterization of Swarming Behavior

Execution of GSO by the individuals in a glowworm swarm gives rise to the emergence of two temporal phases at the group level: (1) splitting of the glowworm swarm into subgroups and (2) local convergence of glowworms in each subgroup to the peak locations. A description of these two phases follows.

3.1.1 Splitting of the Glowworm Swarm into Subgroups

The local decision-domain update rule (refer to (2.4)) enables each glowworm to select its neighbors so that its movements get biased toward the nearest peak. This

© Springer International Publishing AG 2017

K.N. Kaipa and D. Ghose, *Glowworm Swarm Optimization*,

Studies in Computational Intelligence 698, DOI 10.1007/978-3-319-51595-3_3

individual agent behavior leads to a collective behavior of agents that constitutes the automatic splitting of the whole group into subgroups whose number is equal to the number of peak locations and where each subgroup of agents gets allocated to a nearby peak (that is a peak to which the agent-distance, averaged over the subgroup, is minimum among those distances to all the peaks in the environment). The simulation result in Chap. 2 (Fig. 2.8) demonstrated this group-behavior where the swarm splits into three subgroups and movements of agents in each subgroup get biased toward one of the peaks.

In the following, some more simulations are used in order to clearly characterize the splitting behavior of the agent-swarm. In particular, it is shown how the same initial placement of agents gives rise to different splitting behaviors as conditioned by factors like the placements of various peaks, peak values, and slope of the function-profile in the vicinity of the peaks. The $J_2(x, y)$ function (3.1) is considered for this set of experiments (Fig. 3.1).

$$J_2(x, y) = \sum_{i=1}^{Q} a_i \exp(-b_i((x - x_i)^2 + (y - y_i)^2)) \tag{3.1}$$

where, Q represents the number of peaks and (x_i, y_i) represents the location of each peak. The function $J_2(x, y)$ represents a linear sum of two dimensional exponential functions centered at the peak-locations. The constant b_i determines how the function profile changes slope in the vicinity of the peak i. Initially, b_i is kept equal for all the individual exponentials by choosing $b_i = 3, i = 1, \ldots Q$. A workspace of $(-5, 5) \times (-5, 5)$ and a set of ten peaks ($Q = 10$) are considered for the purpose. The values of a_i, x_i, and y_i are generated according to the following equations:

$$a_i = 1 + 2\vartheta$$
$$x_i = -5 + 10\vartheta$$
$$y_i = -5 + 10\vartheta$$

where, ϑ is uniformly distributed within the interval [0, 1].

Note that a_i, x_i, and y_i are random variables and each set of instantiations of these random variables gives rise to a different multimodal function profile. The function profiles generated in the above manner are representative of a varied set of problem scenarios: they have peaks at different random points, giving rise to profiles with closely spaced peaks, distant peaks, and peaks on the edges of the workspace. More-over, generating a_i randomly for each peak gives rise to profiles with equal/unequal peaks. Two different function profiles are generated, each represented by J_2^a and J_2^b. The corresponding values of $\{a_i, x_i, y_i, i = 1, \ldots, 10\}$ for the two function profiles are shown in Tables 3.1 and 3.2. Figure 3.1a, b show the multimodal function pro-files $J_2^a(x, y)$ and $J_2^b(x, y)$. Figure 3.2a, b show the emergence of solution for each function profile when $n = 300$ and $r_s = 5$. The random initial agent placement is kept same for both the simulations. It is observed that when the slopes of the peaks

Table 3.1 Values of a_i, x_i, and y_i used to generate the function profile $J_2^a(x, y)$

	1	2	3	4	5	6	7	8	9	10
a_i	2.616	2.427	1.223	2.366	1.338	2.285	2.325	1.002	1.709	2.985
x_i	2.933	−4.084	3.459	2.066	−4.814	4.381	−1.111	−1.711	−1.728	−3.233
y_i	−0.559	−2.146	−2.803	2.710	4.789	3.282	−1.496	−3.977	0.244	−3.166

Table 3.2 Values of a_i, x_i, and y_i used to generate the function profile $J_2^b(x, y)$

	1	2	3	4	5	6	7	8	9	10
a_i	1.822	1.966	2.957	1.210	2.626	1.977	2.761	2.612	2.748	1.503
x_i	−1.214	−3.807	−4.768	−3.613	4.976	−1.716	−1.751	3.798	2.993	0.135
y_i	3.677	−4.550	−2.056	0.772	4.277	0.856	−2.574	0.088	1.222	0.699

are equal, the swarm splits into subgroups according to the Voronoi-partition of the peak locations (Fig. 3.2a, b), that is, agents deployed in the Voronoi-partition of a peak location remain in the same partition, during their movements, and eventually converge to the corresponding Voronoi-center that coincides with the peak location. However, a few agents located near the border region of a Voronoi-partition migrate into adjacent partitions and eventually get co-located at the respective peaks. Moreover, the same placement of agents, with a change in the location of the peaks, gives rise to a different partitioning of the agents. However, note that the splitting behavior of the swarm respects the Voronoi-partition of the new set of peak locations (Fig. 3.2b). The locations of agents at different time instants ($t = 0, 20, 40, 60$) are plotted in Fig. 3.3a–d, which show the formation of subgroups as a function of time.

Next, the splitting of the agents is characterized in the case where different slope profiles (b_i) are used for the individual exponentials in the $J_2(x, y)$ function. For this purpose, two peaks at $(−3, 0)$ and $(3, 0)$ are considered, with $a_1 = 3$, $b_1 = 0.8$ and $a_2 = 2$, $b_2 = 0.1$, respectively. Figure 3.4 shows the emergence of agent movements when 100 agents are randomly deployed in the workspace. The line L, that passes through each equi-valued contour at a point where the gradient shifts direction from one peak to the other, divides the workspace into two *attraction-regions*. Note that all agents on the concave side of line L converge to the left peak and a majority (86%) of the agents on the convex side of the line L converge to the right peak. When there are more than two peaks, the attraction-regions for each peak can be obtained in a similar manner as described above and it can be shown that the swarm splits into subgroups according to the attraction-regions and eventually converge to the respective peak locations. It is easy to see that the partitioning of the region obtained based on the attraction-regions of the peaks coincides with the Voronoi-partition of the peaks when their slopes become equal.

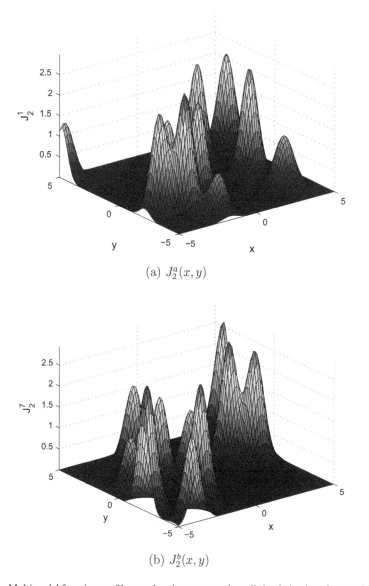

(a) $J_2^a(x, y)$

(b) $J_2^b(x, y)$

Fig. 3.1 Multimodal function profiles used to demonstrate the splitting behavior of agents in GSO

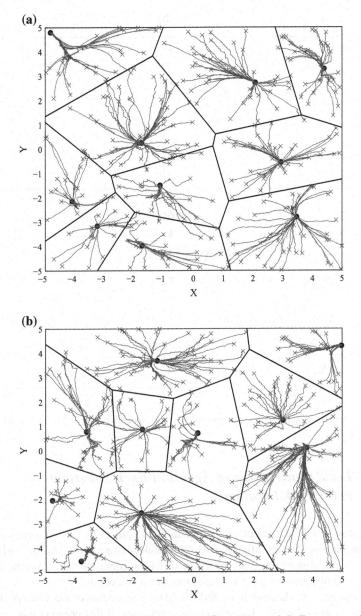

Fig. 3.2 **a** Emergence of agent-movements when $J_2^a(x, y)$ is used. **b** Emergence of agent-movements when $J_2^b(x, y)$ is used. The swarm splits according to the Voronoi-partition of the peak locations in both the cases

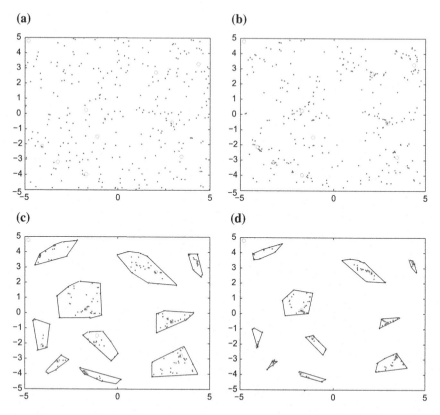

Fig. 3.3 Location of agents at different time instants (corresponding to Fig. 3.2a). **a** $t = 0$. **b** $t = 20$. **c** $t = 40$. **d** $t = 60$

3.1.2   Local Convergence of Agents in Each Subgroup to the Peak Locations

The relative initial placement of the agents with respect to various peaks in the environment gives rise to different subgroup-peak configurations. Accordingly, the respective local convergence behaviors are different from each other. Among these, two major configurations are considered that occur frequently:

Configuration 1: The peak is located within the convex-hull of the initial positions of agents in the subgroup. The convergence behavior of this configuration can be explained in the following way. For simplicity, a radially symmetric function-profile with a single peak at the center and an initial placement of three agents a, b, and c as shown in Fig. 3.5a are considered. At time instant t_p, agent a remains stationary, agent b makes a deterministic movement toward a, and agent c moves either toward a or b (since, $\ell_a > \ell_b > \ell_c$). Note from Fig. 3.5b that the agent movements at any time instant are within the convex hull of all the current positions of agents. Agent a does

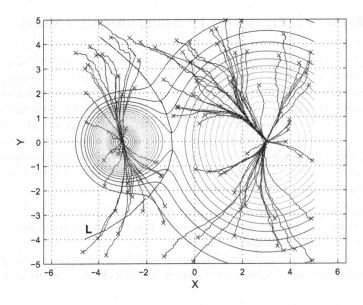

Fig. 3.4 Emergence of agent-movements when two unequal peaks with different slopes are considered. The line L divides the region into two *attraction-regions*

not move until after two time steps (i.e., at $t_p + 2$), when b would have crossed the equi-valued contour $C_a(t_p + 2)$, leading to the condition $\ell_b > \ell_a$. Now, b remains stationary and a starts moving toward b. This cycle repeats, leading to the asymptotic convergence of agents to the peak.

Configuration 2: The peak is located outside the convex-hull of initial agent positions and all the agents are situated on one side of the peak. In order to describe the convergence behavior, the same radially symmetric function-profile with a single peak at the center is considered. However, an initial placement of agents a, b, and c such that b and c are located on one side of the tangent line T at agent a's location is considered as shown in Fig. 3.6a. At time instant t_p, agent a remains stationary, agent b makes a deterministic movement toward a, and agent c moves either toward a or b (since, $\ell_a > \ell_b > \ell_c$). Agent a does not move until after two time steps (i.e., at $t_p + 2$), when b leapfrogs over a, leading to the condition $\ell_b > \ell_a$ (Fig. 3.6b). Unlike in the previous case, agents movements at any instant are not necessarily within the convex-hull of the current agent positions. For example, at $t_p + 2$, agent a moves outside the convex-hull when it leapfrogs over b. However, the leapfrogging mechanism ensures that the agents eventually converge to the peak.

3.2 Theoretical Framework for a Simplified GSO Model

The analysis of GSO is restricted to the local convergence of agents to a *leader* (to be defined more precisely later). Leaders may represent agents that are closer to the peaks than other agents. For instance, in Figs. 3.5 and 3.6, agent *a* is the leader. Note that multiple peaks may give rise to multiple leaders. From the above analysis of local convergence behavior of the agents in the vicinity of a peak, it is clear that the leader does not move until another agent crosses the equi-valued contour that passes through the leader's position in the first case, and until another agent leapfrogs over the leader as in the second case. The simplified model is obtained by making the following modifications to the actual GSO algorithm:

Modification 3.1   Each agent i contains a constant luciferin value ℓ_i.

Modification 3.2   The local-decision domain range is kept constant and is made equal to the maximum sensing range r_s.

Modification 3.3   The step-size s is modified such that an agent reaches the leader's location in one step, whenever it is situated within a distance s to the leader.

Fig. 3.5 a State of agents in configuration 1 at time t_p. **b** State of agents in configuration 1 at time $t_p + 2$

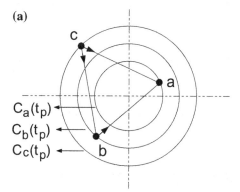

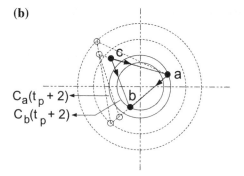

Fig. 3.6 a State of agents in configuration 2 at time t_p. **b** State of agents in configuration 2 at time $t_p + 2$

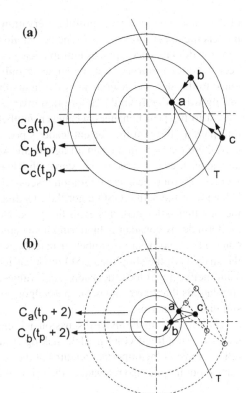

The above assumptions are reasonable within the theoretical framework of local convergence to leaders that is considered here and can be justified as follows. Multimodal function profiles with a property that the inter-peak distance is more than r_s and that the function value in the local vicinity of each peak monotonically decreases with distance from the peak in all directions, are considered. However, the function need not be radially symmetric at each peak. Note that during agent-movements in such a local region, even though the absolute values of agents' luciferin levels change with time, the relative order of luciferin levels of the agents in the same subgroup do not change, which is also reflected by the constant luciferin case used for our analysis. Usually, when multiple leaders (corresponding to multiple peaks) are located with inter-leader distances of the order of r_s, a typical situation occurs where an agent is located in the r_s-neighborhoods of, and hence under the influence of, all the leaders. The primary advantage of using an adaptive local-decision domain is to enable an agent to select its neighbors such that its movements get biased toward a nearby leader leading to its co-location with it in finite time. Keeping the range of the decision domain constant modifies the above behavior in the sense that each leader has an influence on the agent, which is a function of its associated luciferin level as well as its distance to the agent, leading to the asymptotic convergence of the agent to one of the leaders. However, a careful observation will reveal that this assumption still models the agent movements in the actual algorithm during the period when local-decision domain encompasses multiple leaders and until a stage is reached when the decision domain shrinks and encompasses only the nearest leader. The modification made to the step-size s influences the first subgroup-peak configuration in the following way. Note from the description given earlier that the leader (agent which is closest to the peak) changes whenever another agent crosses the equi-valued contour passing through the current leader's position. However, with the modification of s, there exists a time when all agents are within s-distance to the current leader and hence they converge in one step to that leader. The modification in s affects the second subgroup-peak configuration discussed above by removing the leapfrogging behavior of the actual GSO algorithm, leading to the convergence of the agents to the one that is located closest to the peak. Note that the algorithmic constants − luciferin decay constant ρ, luciferin-enhancement constant γ, decision-range update constant β, and desired number of neighbors n_t − need not be considered during the analysis of the simplified GSO model as it is assumed that agents maintain constant luciferin and local-decision range values. However, in Chap. 4, it is shown in simulations that fixed values for these algorithmic constants work well for different simulation scenarios.

Therefore, from the above discussion, it is clear that while the above modifications lead to a simplified GSO model, making it amenable to analysis, the model still reflects most of the important features of the actual algorithm and hence the analysis carried out here is also relevant to the original GSO algorithm.

3.2.1 Notations

$I = \{1, 2, \ldots, n\}$ is the index set used to identify the n agents (glowworms).
$x = (x_1, \ldots, x_n)$ is the state vector containing the states (positions) of all members of the group.
$d_{ij} = \|x_i - x_j\|$ is the Euclidean distance between locations of agents i and j.
r_s is the radial sensor range of each agent.
ℓ_i is the *luciferin*-value associated with agent i.
$N_i = \{j : j \neq i, j \in I, d_{ij} \leq r_s\}$ is the set of distance-based neighbors of agent i.
$H_i = \{j : j \in I \text{ and } \ell_i < \ell_j\}$ is the set of luciferin-based neighbors of agent i.
$L_i = N_i \cap H_i$ is the leader-set of agent i.
t_k is the k-th discrete time instant.
$G(V, E)$ is the directed graph with the set of nodes (glowworms) $V = \{v_1, v_2, \cdots v_n\}$ and the edge set $E \subseteq V \times V$ (an edge (i, j) exists iff $j \in L_i$).
$L = \{i : i \in I \text{ and } L_i(t_k) = \emptyset, \text{ for all } t_k\}$ is the Leader-set of G.
$F = I - L$ is the follower-set of G.
$\top = \{0, 1, \cdots, k, k + 1, \cdots\}$ is the discrete time-index set.
In the following, the terms glowworm, node, and agent are used interchangeably.

3.2.2 Assumptions

Assumption 3.1 Each agent is represented by a point.

Assumption 3.2 Agents move instantaneously with a step-size of s at each iteration.

Assumption 3.3 Collisions between the agents are ignored.

Assumption 3.4 The value of ℓ_j remains constant, for all $j \in I$.

Assumption 3.5 The range of local-decision domain of each agent is kept constant and made equal to the maximum sensor range, i.e., $r_d^i = r_s$.

Note: Assumption 3.4 is related to Modification 3.1 and Assumption 3.5 is related to Modification 3.2 discussed earlier.

3.2.3 GSO Model for Analysis

The continuous-time model of the glowworm dynamics can be stated as:

$$\dot{x}_i = u_i \tag{3.2}$$

$$u_i = \begin{cases} 0, & \text{if } L_i = \emptyset \\ a_{ij}(x_j - x_i), & \text{otherwise} \end{cases} \tag{3.3}$$

where, $j \in L_i$, i selects to move toward j with a probability p_j given by (2.2), and a_{ij} is the movement-gain. In particular, the discrete-time model of the agent dynamics given in (2.3) is restated below:

$$x_i(t_{k+1}) = x_i(t_k) + s \left(\frac{x_j(t_k) - x_i(t_k)}{\|x_j(t_k) - x_i(t_k)\|} \right) \tag{3.4}$$

where

$$s = \begin{cases} \delta \text{ if } d_{ij}(t_k) \geq \delta \\ d_{ij}(t_k), & \text{otherwise} \end{cases} \tag{3.5}$$

where, $0 < \delta < r_s$.

The above definition of step-size s incorporates Modification 3.3 discussed earlier.

Some definitions and lemmas are provided before presenting the main results of the chapter.

Definition 3.1 An agent i is said to be stationary at time t_k, if $x_i(t_{k+1}) = x_i(t_k)$.

Definition 3.2 An agent i is said to be stationary for all time starting from t_j, if $x_i(t_{k+1}) = x_i(t_k)$, for all $k \geq j$.

Definition 3.3 $G(V, E)$ is said to be stationary at time t_k, if all the agents are stationary at time t_k, i.e., $x_i(t_{k+1}) = x_i(t_k)$, for all $i \in I$.

Definition 3.4 $G(V, E)$ is said to be stationary for all time after t_j, if all the agents are stationary for all time after t_j, i.e., $x_i(t_{k+1}) = x_i(t_k)$, for all $k \geq j$ and for all $i \in I$.

Definition 3.5 Two nodes i and j are co-located at time t_k if $x_i(t_k) = x_j(t_k)$ or $d_{ij}(t_k) = 0$.

Definition 3.6 A leader $l_i \in L$ is said to be *isolated*, if $\|l_i - l_j\| > 3r_s$ for all $j \in L$, $j \neq i$.

Lemma 3.1 *At least one agent remains stationary always.*

Proof From Assumption 3.4, there exists $i^* \in I$ such that $\ell_{i^*}(t_k) \geq \ell_i(t_k)$, for all $i \in I$ and $i \neq i^*$ and for all t_k. Therefore, $H_{i^*}(t_k) = \emptyset$, for all t_k. This implies that $L_{i^*}(t_k) = N_{i^*}(t_k) \cap H_{i^*}(t_k) = \emptyset$, for all t_k. This gives $u_{i^*}(t_k) = 0$, for all t_k. From (3.4), $x_{i^*}(t_{k+1}) = x_{i^*}(t_k)$, for all t_k. Therefore, the agent i^* remains stationary for all time t_k. $\qquad\square$

Note that Assumption 3.5 is not required to prove the above result.

Lemma 3.2 *An agent with an empty leader set at some time t_k is stationary at t_k.*

Proof The proof is straightforward and hence is omitted. $\qquad\square$

Lemma 3.3 *If G is stationary at time t_q, then it is stationary for all $t_k, k > q$.*

Proof The result is proved using mathematical induction. Let G be stationary at time t_q. From Definition 3.1, $x_i(t_{q+1}) = x_i(t_q)$ for all $i \in I$. Therefore, the result is true for $k = q + 1$. Now, let $x_i(t_{k+1}) = x_i(t_k)$, for all $i \in I$ and for some $k > q$. From (3.3)–(3.5), this is true, if and only if for all i, either
(i) $L_i(t_k) = \emptyset$ or
(ii) $L_i(t_k) \neq \emptyset$ and $d_{ij}(t_k) = 0$, for all $j \in L_i(t_k)$.
Suppose (i) is true. Since there are no agent movements at t_k, the leader-sets continue to be empty at t_{k+1}, i.e., $L_i(t_{k+1}) = \emptyset$, for all $i \in I$. From (3.2) and (3.3), we get $x_i(t_{k+2}) = x_i(t_{k+1})$, for all $i \in I$.
Suppose (ii) is true. Since there are no agent movements, $L_i(t_{k+1}) = L_i(t_k) \neq \emptyset \Rightarrow d_{ij}(t_{k+1}) = d_{ij}(t_k) = 0$, for all $j \in L_i(t_{k+1})$. Using (3.3)–(3.5) again, we get $x_i(t_{k+2}) = x_i(t_{k+1})$. It is proved that if G is stationary at time $t_k, k > q$, then it is stationary at t_{k+1}. Thus, the result follows by induction. $\qquad\square$

Lemma 3.4 *For any agent i, if $L_i(t_k) \neq \emptyset$ at some time t_k, then $L_i(t_k) \neq \emptyset$, for all $q \geq k$.*

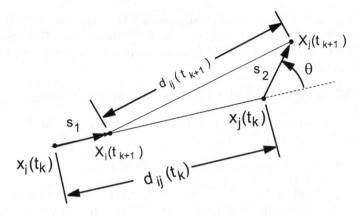

Fig. 3.7 Movements of agents i and j at iteration t_k

Proof Given $L_i(t_k) \neq \emptyset$. Let $j \in L_i(t_k)$ such that i selects to move toward it. This implies that $d_{ij}(t_k) \leq r_s$. Agent i moves a distance s_1 units on the line joining $x_i(t_k)$ and $x_j(t_k)$. During the same time, let j move s_2 units in a direction that makes an angle θ with the vector $x_j(t_k) - x_i(t_k)$ (see Fig. 3.7).

Case 1: Suppose $s_1 = \delta$ and $s_2 \leq \delta$ Using the triangle inequality, we have

$$
\begin{aligned}
d_{ij}(t_{k+1}) &\leq (d_{ij}(t_k) - s_1) + s_2 \\
&= (d_{ij}(t_k)) - (s_1 - s_2) \\
&\leq d_{ij}(t_k)(\text{since} s_1 - s_2 \geq 0) \\
&\leq r_s
\end{aligned}
\tag{3.6}
$$

Case 2: Suppose $0 < s_1 < \delta$ and $s_2 = \delta$.

This implies, i reaches j's position at t_k in one step, i.e., $x_i(t_{k+1}) = x_j(t_k)$ while j moves a step distance of δ units. Therefore, $d_{ij}(t_{k+1}) = \delta$. From Assumption 3.2 and (3.6), we get $L_i(t_{k+1}) \neq \emptyset$. Similarly, it can be shown that, if $L_i(t_q) \neq \emptyset$ for some $q > k$, then $L_i(t_{q+1}) \neq \emptyset$. Thus, the desired result is obtained by induction. □

Lemma 3.4 implies that once an agent acquires at least one leader, it continues to have a leader.

Lemma 3.5 *Suppose G becomes stationary at some time t_k. If nodes i and j are not co-located and $\ell_i \neq \ell_j$, then $\|x_i(t_k) - x_j(t_k)\| > r_s$.*

Proof Since the values of ℓ are distinct, let $\ell_i < \ell_j \Rightarrow j \in H_i(t_k)$. Now, $x_i(t_{k+1}) = x_i(t_k)$ (since G becomes stationary at time t_k) and $d_{ij}(t_k) \neq 0 \Rightarrow j \notin L_i(t_k)$ (otherwise i can move toward j). This is true only if $j \notin N_i(t_k)$ (since $L_i = N_i \cap H_i$ and $j \in H_i$). □

Lemma 3.6 *If for a node i, $L_i(t_q) \neq \emptyset$ at any time t_q, then when node i becomes stationary at t_k, there exists at least one j co-located with i such that $\ell_j > \ell_i$.*

Proof $L_i(t_q) \neq \emptyset \Rightarrow L_i(t_k) \neq \emptyset$ (from Lemma 3.4). Suppose $d_{ij}(t_k) \neq 0$, for all $j \in L_i(t_k)$. From (3.3), $x_i(t_{k+1}) \neq x_i(t_k)$, which is a contradiction. Therefore, there exists at least one j co-located with i when it becomes stationary such that $\ell_i < \ell_j$. Hence the result. □

Lemma 3.7 *At any time t_k, let $\hat{L}(t_k) = \{i : L_i(t_k) = \emptyset\}$, then $L \subseteq \hat{L}(t_k)$.*

Proof L is the leader-set of G, i.e., if $i \in L$ then $L_i(t_k) = \emptyset$, for all $t_k \Rightarrow i \in \hat{L}(t_k)$, for all $i \in L \Rightarrow L \subseteq \hat{L}(t_k)$. □

Lemma 3.8 *Let $\hat{L}$ be defined as in Lemma 3.7. Then, $\hat{L}(t_{k+1}) \subseteq \hat{L}(t_k)$.*

Proof For $i \in I$ and $i \notin \hat{L}(t_k)$, $L_i(t_k) \neq \emptyset \Rightarrow L_i(t_{k+1}) \neq \emptyset$ (from Lemma 3.4) $\Rightarrow i \notin \hat{L}(t_{k+1})$. This implies that $i \in \hat{L}(t_{k+1}) \Rightarrow i \in \hat{L}(t_k)$. Hence we get the result. □

Lemma 3.8 implies that a non-member of $\hat{L}(t_k)$ cannot become a member of $\hat{L}(t_{k+1})$. However, a member $i \in \hat{L}(t_k)$ may lose its membership in $\hat{L}(t_{k+1})$. For instance, this happens when an agent, say j, with $\ell_i < \ell_j$ enters within the r_s range of i at t_{k+1}.

Lemma 3.9 *There exists a time t_q so that $\hat{L}(t_k) = \hat{L}(t_{k+1})$, for all $k \geq q$.*

Proof The relation $\hat{L}(t_{k+1}) \subset \hat{L}(t_k)$ is true only if there exists at least one i such that $i \in \hat{L}(t_k)$, but $i \notin \hat{L}(t_{k+1})$ (this occurs when i acquires a leader). Since $L \subseteq \hat{L}(t_{k+1}) \subseteq \hat{L}(t_k)$ (from Lemmas 3.6 and 3.7), members can continue to lose membership in $\hat{L}$ only until t_q when $\hat{L}(t_q) = L$. Therefore, for all $t_k, k \geq q, \hat{L}(t_k) = \hat{L}(t_{k+1})$. □

3.3 Main Results: Constant Luciferin Case

The movements of followers are analyzed when they are influenced by the presence of multiple leaders. For simplicity, only two leaders $l_1, l_2 \in L$ are considered. Based on the distance between the two leaders, the analysis of follower movements and convergence to leader-locations can be classified into three cases (Fig. 3.8), which are described as follows:

- *Isolated neighborhoods:* $\|x_{l_1} - x_{l_2}\| > 3r_s$; all agents within the r_s-neighborhood of l_1 are attracted only toward l_1 (Fig. 3.8a). The isolated nature can be understood by considering the case $\|x_{l_1} - x_{l_2}\| = 3r_s$ and two agents a and b located on the intersection points of the line joining the two leader positions and the neighborhood boundaries of the two leaders, respectively (Figure 3.8b). Now, agent a is influenced by l_1 as well as b. Therefore, whenever $\|x_{l_1} - x_{l_2}\| > 3r_s$, all agents in the neighborhood of a leader are influenced only by that leader and isolated from agents located in other leader-neighborhoods. It is proved that all the followers converge to the isolated leader's location in finite time (Theorem 3.1).

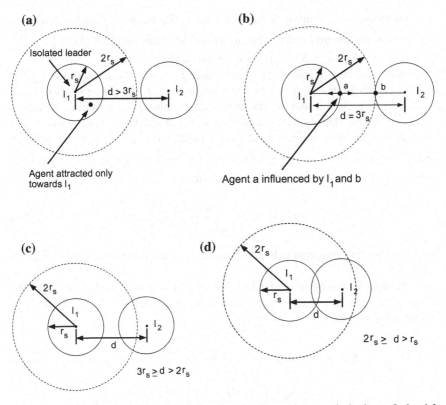

Fig. 3.8 Different cases of neighborhoods based on the distance between the leaders: **a** Isolated. **b** The case when $d = 3r_s$. **c** Not isolated and non-overlapping. **d** Overlapping

- *Non-isolated and non-overlapping neighborhoods*: $2r_s < \|x_{l_1} - x_{l_2}\| \le 3r_s$; An agent within r_s-neighborhood of l_1 can possibly be attracted toward an agent within r_s-neighborhood of l_2 (Fig. 3.8c). It is proved that followers converge to either one of the leader locations in finite time (Theorem 3.2).
- *Overlapping neighborhoods*: $r_s < \|x_{l_1} - x_{l_2}\| \le 2r_s$ (Fig. 3.8d). It is proved that the agents located in the intersection of leader-neighborhoods asymptotically reach either one of the leader locations, while the remaining agents do so in finite time (Theorem 3.3).

Theorem 3.1 *If a leader l is isolated (Definition 3.6), then all the members of $N_l(t_0)$ are co-located with l, for all $t_k, k \ge K(q)$, where, $q = |F|$, $K(i) = \sum_{j=1}^{i} \left\lceil \frac{d_{f_j l}(t_{K(j-1)})}{\delta} \right\rceil$, $K(0) = 0$, $f_j \in F$, $j = 1, \ldots, q$, and $\ell_{f_1} \ge \ell_{f_2} \ge \cdots \ge \ell_{f_q}$.*

Proof Consider an isolated leader l. Since $\ell_l > \ell_{f_j}$ for all $f_j \in N_l(t_0)$, all followers are influenced by the leader. Since the values of ℓ are constant, the followers can be sorted as $f_1, f_2, \ldots, f_q$ according to ascending order of their associated ℓ values. Therefore, $\ell_{l_1} > \ell_{f_1} \ge \ell_{f_2} \ge \cdots \ge \ell_{f_q}$.

Consider the movements of f_1. Since $L_{f_1}(t_k) = \{l\}$, for all t_k, f_1 makes deterministic movements toward l and reaches it in $t_{K(1)}$ iterations, where $K(1) = \left\lceil \frac{d_{f_1l}(0)}{\delta} \right\rceil$. Therefore, for all $t_k, k \geq K(1)$, f_1 is co-located with l.

Consider the movements of f_2. Since either $L_{f_2}(t_k) = \{l\}$ or $L_{f_2}(t_k) = \{l, f_1\}$, for all t_k, at each iteration it moves either toward l or f_1. Based on its distance to l at t_0 and its moves, it may or may not reach l within $t_{K(1)}$ iterations. However, at $t_{K(1)}$ (assuming that f_2 is not co-located with l at that time), since l and f_1 are co-located, f_2 has only one direction to move. Let $d_{f_2l}(t_{K(1)})$ be the distance between f_2 and l at $t_{K(1)}$. Then f_2 converges to l in $\left\lceil \frac{d_{f_2l}(t_{K(1)})}{\delta} \right\rceil$ steps after $t_{K(1)}$. Therefore, for all $t_k, k \geq K(2)$, f_2 is co-located with l, where

$$K(2) = K(1) + \left\lceil \frac{d_{f_2l}(t_{K(1)})}{\delta} \right\rceil \tag{3.7}$$

Note that this is a conservative bound because f_2 could get co-located with l much before $t_{K(2)}$.

Now, suppose f_i is co-located with l, for all $t_k, k \geq K(i)$. Let $d_{f_{i+1}l}(t_{K(i)})$ be the distance between f_{i+1} and l at time $t_{K(i)}$. Now, f_{i+1} takes $\left\lceil d_{f_{i+1}l}(t_{K(i)}) \right\rceil$ steps to reach l. This implies,

$$K(i+1) = K(i) + \left\lceil \frac{d_{f_{i+1}l}(t_{K(i)})}{\delta} \right\rceil$$

$$= \sum_{j=1}^{i} \left\lceil \frac{d_{f_jl}(t_{K(j-1)})}{\delta} \right\rceil + \left\lceil \frac{d_{f_{i+1}l}(t_{K(i)})}{\delta} \right\rceil$$

$$= \sum_{j=1}^{i+1} \left\lceil \frac{d_{f_jl}(t_{K(j-1)})}{\delta} \right\rceil$$

Therefore, by induction, it can be shown that all members of $N_l(t_0)$ will be co-located with l, for all $t_k, k \geq K(q)$. $\square$

Theorem 3.2 *Let K be defined as in Theorem 3.1 and $q = |F|$. Suppose two leaders l_1 and l_2 are not isolated but their neighborhoods are non-overlapping, i.e., $2r_s < \|l_1 - l_2\| \leq 3r_s$, then all the followers get co-located with either one of the leaders, for all $t_k, k \geq K(q)$.*

Proof Consider followers $f_1, f_2, \ldots, f_q$ such that $\ell_{l_1} > \ell_{f_1} \geq \cdots \geq \ell_{f_g} > \ell_{l_2} > \ell_{f_{g+1}} \geq \cdots \geq \ell_{f_q}$.

Consider the movements of f_1. Obviously, $f_1 \in N_{l_1} \bigcup N_{l_2}$. Otherwise, there will be no attracting influence on f_1 and hence it remains stationary, which then contradicts the characteristics of a follower. Moreover, $f_1 \notin N_{l_2}$ because the relationship $\ell_{f_1} > \ell_{l_2}$ contradicts the fact that a leader must have maximum ℓ-value in its neighborhood. Therefore, $f_1 \in N_{l_1}(t_k)$, for all k. This implies that f_1 gets co-located with l_1, for all $t_k, k \geq K(1)$.

Consider the movements of f_2. Clearly, $f_2 \notin N_{l_2}$. Also, $f_2 \notin N_{l_1} \bigcup N_{l_2}$ is possible for t_k such that $k \le K(1)$. However, during this time $L_{f_2} = \{f_1\}$. Otherwise, f_2 becomes a leader which is a contradiction. But once f_1 is co-located with l_1, f_2 should be within r_s-distance of l_1. Therefore, for all $t_k, k \ge K(2)$, f_2 is co-located with l_1. Similar analysis can be carried out till follower f_g and it can be shown that $f_1, f_2, \ldots, f_g$ will be co-located with l_1 for all $t_k, k \ge K(g)$.

Consider the movements of f_{g+1}. If $f_{g+1} \in N_{l_1}(t_0)$, it remains in N_{l_1} and gets co-located for all $t_k, k \ge K(g+1)$. If $f_{g+1} \in N_{l_2}(t_0)$, there is a possibility that it leaves N_{l_2} and enters N_{l_1} before $f_1, \ldots, f_g$ reach l_1. However, once it enters N_{l_1}, it cannot leave N_{l_1}. After $t_{K(g)}$, $f_{g+1} \in N_{l_1} \bigcup N_{l_2}$. Therefore, if $f_{g+1} \in N_{l_1}$ (or $f_{g+1} \in N_{l_2}$), it gets co-located with l_1 (or l_2), for all $t_k, k \ge K(g+1)$. Note that while evaluating the expression $\left\lceil \frac{d_{f_j l}(t_{K(j-1)})}{\delta} \right\rceil, l = l_1$ (or l_2) if $f_j \in N_{l_1}(t_{K(j-1)})$ (or $f_j \in N_{l_2}(t_{K(j-1)})$). In general, if $f_i \in \{f_{g+1}, \ldots, f_q\}$ reaches either one of the leaders in time $t_{K(i)}$, then f_{i+1} reaches either one of them in time $t_{K(i+1)}$. Therefore, all the followers $f_1, \ldots, f_g$ are co-located at l_1, for all $t_k, k \ge K(g)$ and the rest of the followers $f_{g+1}, \ldots, f_q$ are co-located at either one of the leaders for all $t_k, k \ge K(q)$. $\qquad\square$

Theorem 3.3 *Let K be defined as in Theorem 3 and $q = |F|$. If two leaders l_1 and l_2 have overlapping neighborhoods, i.e., $r_s < \|l_1 - l_2\| \le 2r_s$, then all followers f_j such that $j \in F$ and $\ell_{l_2} < \ell_{f_j} < \ell_{l_1}$ converge to one of the leaders in finite time and the remaining followers located in the overlap region of the neighborhoods asymptotically converge to one of the leaders.*

Proof Let the followers $f_1, f_2, \ldots, f_q$ satisfy $\ell_{l_1} > \ell_{f_1} \ge \cdots \ge \ell_{f_g} > \ell_{l_2} > \ell_{f_{g+1}} \ge \cdots \ge \ell_{f_q}$. Consider the movements of the members of $F_1 = \{f_1, \ldots, f_g\}$. The set F_1 has the property that $f_i \in N_{l_1} \backslash (N_{l_1} \cap N_{l_2})$, for all $f_i \in F_1$. Using Theorem 3.1, it can be shown that all members of F_1 are co-located with l_1, for all $t_k, k \ge K(g)$.

Consider the movements of f_{g+1} after $t_{K(g)}$. If $f_{g+1} \in N_{l_1} \backslash (N_{l_1} \cap N_{l_2})$, then f_{g+1} reaches l_1 in time $t_K(g+1)$, with $l = l_1$. If $f_{g+1} \in N_{l_2} \backslash (N_{l_1} \cap N_{l_2})$, then f_{g+1} reaches l_2 in time $t_K(g+1)$, with $l = l_2$.

Suppose $f_{g+1} \in N_{l_1} \cap N_{l_2}$.

Let $\Gamma = \{x : x \in (\alpha x_a + (1-\alpha)x_b), 0 \le \alpha \le 1\}$, where x_a and x_b are the points of intersection of the line joining l_1 and l_2 with circles centered at x_{l_1} and x_{l_2}, respectively. Based on the position of f_{g+1} with respect to the line Γ, two cases are obtained as shown in Fig. 3.9.

Case 1: $f_{g+1} \in \Gamma$.

Let p_1 and p_2 ($p_1 + p_2 = 1$) be the probabilities that f_{g+1} makes a step movement toward l_1 and l_2, respectively, at each time t_k. Let $d_{l_1 l_2}$ be the distance between the two leaders' positions. Since the agent moves in discrete steps, the convergence problem can be formulated as a finite state Markov chain with $n = \lceil 2r_s - d_{l_1 l_2} \rceil$, where the agent's position at any iteration t_k is represented by one of the states. For instance, in Fig. 3.10, x_a coincides with State 1 and x_b coincides with State n. Note that the movements of the follower on the line segments $l_1 x_a$ and $l_2 x_b$ need not be considered because, once the agent reaches x_a (x_b) it will be influenced by l_1 (l_2) only, thus

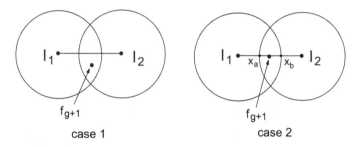

Fig. 3.9 The two different cases that occur when $f_{g+1} \in N_{l_1} \cap N_{l_2}$

Fig. 3.10 The n-state finite
state Markov chain

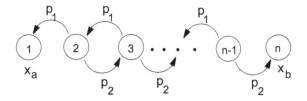

reaching it in finite number of steps. The one-step probabilities can be expressed in
the form of a transition probability matrix given by:

$$P = [p_{ij}]_{n \times n} = \begin{bmatrix} 1 & 0 & . & . & . & 0 \\ p_1 & 0 & p_2 & 0 & . & 0 \\ 0 & p_1 & 0 & p_2 & . & 0 \\ . & . & . & . & . & . \\ 0 & . & . & p_1 & 0 & p_2 \\ 0 & 0 & 0 & 0 & 0 & 1 \end{bmatrix} \qquad (3.8)$$

Let $P^n = [p_{ij}^n]$ be the n-state transition matrix where p_{ij}^n represents the probability
that an agent, initially in state i, ends up in state j at the nth step. The steady state
probability state vector equation is then given by:

$$P^\infty P = P^\infty \qquad (3.9)$$

where $P^\infty = \lim_{n \to \infty} [p_{ij}^n]$ Let

$$P^\infty = \begin{bmatrix} 1 & 0 & \ldots & 0 \\ p_{21} & p_{22} & \cdots & p_{2n} \\ . & . & \cdots & . \\ . & . & \cdots & . \\ p_{(n-1)1} & p_{(n-1)2} & \cdots & p_{(n-1)n} \\ 0 & 0 & \ldots & 1 \end{bmatrix} \qquad (3.10)$$

Consider $P_i^\infty P = P_i^\infty$, where P_i^∞ is the ith row of P^∞. Then,

$$p_1 p_{i2} + p_{i1} = p_{i1};$$
$$p_1 p_{i3} = p_{i2}$$
$$p_1 p_{i4} + p_2 p_{i2} = p_{i3}$$
$$\vdots$$
$$p_1 p_{i(n-1)} + p_2 p_{i(n-3)} = p_{i(n-2)}$$
$$p_2 p_{i(n-2)} = p_{i(n-1)}$$
$$p_{in} + p_2 p_{i(n-1)} = p_{in} \tag{3.11}$$

From (3.11), we get

$$p_{i2} = 0; \quad p_{i3} = 0$$
$$p_1 p_{i(j+2)} + p_2 p_{ij} = p_{i(j+1)}, \, j = 2, 3, \cdots, (n-3)$$
$$p_2 p_{i(n-2)} = p_{i(n-1)}; \quad p_{in} + p_2 p_{i(n-1)} = p_{in} \tag{3.12}$$

From the set of equations in (3.12), we get the following relations:

$$p_{ij} = 0 \text{ for } j = 2, 3, \cdots, (n-1)$$
$$\sum_{j=1}^{n} p_{ij} = 1, \text{ for all } i = 1, 2, \cdots, n$$
$$\Rightarrow p_{i1} + p_{in} = 1, \text{ for all } i = 1, 2, \cdots, n \tag{3.13}$$

However, the set of Eq. (3.13) does not give the complete solution to (3.11). The result in (3.13) shows that the agent starting in any one of the n states, converges to either x_a or x_b asymptotically with probability 1. Once f_{g+1} reaches x_a (x_b), it reaches l_1 (l_2) in $\lceil d_{l_1 l_2} - r_s \rceil$ time steps.

As an illustration, the explicit probabilities of reaching either l_1 or l_2 from any intermediate state when $n = 5$ is given below:

$$P_5^\infty = \begin{bmatrix} 1 & 0 \, 0 \, 0 & 0 \\ \frac{p_1(1-p_1 p_2)}{1-2p_1 p_2} & 0 \, 0 \, 0 & \frac{p_2^3}{1-2p_1 p_2} \\ \frac{p_1^2}{1-2p_1 p_2} & 0 \, 0 \, 0 & \frac{p_2^2}{1-2p_1 p_2} \\ \frac{p_1^3}{1-2p_1 p_2} & 0 \, 0 \, 0 & \frac{p_2(1-p_1 p_2)}{1-2p_1 p_2} \\ 0 & 0 \, 0 \, 0 & 1 \end{bmatrix} \tag{3.14}$$

Fig. 3.11 Follower
movements when
$f_{g+1} \in N_{l_1} \cap N_{l_2} \setminus P$

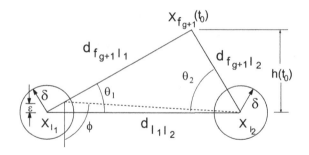

Case 2: $f_{g+1} \in N_{l_1} \cap N_{l_2} \setminus \Gamma$.
At each iteration, f_{g+1} makes a step movement δ toward either l_1 or l_2. Let $d_{l_1 l_2}$ be
the distance between l_1 and l_2. Let $h(t)$ and $h(t+1)$ be the perpendicular distances
from $x_{f_{g+1}}(t)$ and $x_{f_{g+1}}(t+1)$, respectively, to the line joining x_{l_1} and x_{l_2}. Clearly,
$h(t+1) < h(t)$. Therefore, $h(t)$ is a monotonically decreasing function. Now, it is
shown that there exists a time t_q such that

$$d_{l_1 l_2} < d_{f_{g+1} l_1}(t_k) + d_{f_{g+1} l_2}(t_k) < d_{l_1 l_2} + \delta, \text{ for all } t_k, k \geq q \qquad (3.15)$$

Referring to Fig. 3.11, the angles θ_1 and θ_2 are given by:

$$\theta_1 = \arccos\left(\frac{d_{f_{g+1} l_1}^2 + d_{l_1 l_2}^2 - d_{f_{g+1} l_2}^2}{2 d_{f_{g+1} l_1} d_{l_1 l_2}}\right) \qquad (3.16)$$

$$\theta_2 = \arccos\left(\frac{d_{f_{g+1} l_2}^2 + d_{l_1 l_2}^2 - d_{f_{g+1} l_1}^2}{2 d_{f_{g+1} l_2} d_{l_1 l_2}}\right) \qquad (3.17)$$

A lower bound on the perpendicular distance to the line joining l_1 and l_2 such that
$d_{f_{g+1} l_1}(t_k) + d_{f_{g+1} l_2}(t_k) = d_{l_1 l_2} + \delta$ can be calculated as:

$$\epsilon = \frac{\delta(\delta + 2 d_{l_1 l_2})}{2(d_{l_1 l_2} + \delta)} \qquad (3.18)$$

Using (3.18), the minimum perpendicular distance d_p moved by f_{g+1} at each step is
given by:

$$d_p = \delta \cos \phi$$

where,   (3.19)

$$\phi = \frac{\pi}{2} - \arctan\left(\frac{\epsilon \tan \theta}{d \tan \theta - \epsilon}\right)$$

$$\theta = \min\{\theta_1, \theta_2\}$$

From Fig. 3.11, the initial perpendicular distance $h(t_0)$ is given by:

$$h(t_0) = \frac{d_{l_1 l_2} \tan \theta_1 \tan \theta_2}{\tan \theta_1 + \tan \theta_2} \tag{3.20}$$

Therefore, after t_q, $q = \left\lceil \frac{h(t_0) - \epsilon}{d_p} \right\rceil$, the perpendicular distance to the line segment $l_1 l_2$ is within ϵ and hence (3.15) is satisfied. The consequence of (3.15) is that

$$\left\lceil d_{f_{g+1} l_1}(t_k) + d_{f_{g+1} l_2}(t_k) \right\rceil = \left\lceil d_{l_1 l_2} \right\rceil, \text{ for all } k \geq q$$

Hence, a similar analysis as in Case 1 can be used to prove asymptotic convergence to one of the leaders. Moreover, since the movement of f_{g+1} is always on the line joining its own position and that of l_1 or l_2, it cannot converge to a position between the leader positions and gets co-located with either l_1 or l_2. A similar analysis can be done to show asymptotic convergence of the remaining followers $(f_{g+2}, \ldots, f_q)$ to either one of the leaders. □

Remarks: It can be inferred from the above theoretical results that, even though agent movements are probabilistic in nature, agents get co-located with leaders in finite time if the leader-neighborhoods are non-overlapping. Otherwise, the agents get co-located with leaders asymptotically.

3.4 Results for the Variable Luciferin Case

In all the previous theorems, it was assumed that the agents have constant luciferin values. However, the luciferin values of agents in the actual GSO algorithm vary with time and their position in the workspace. Therefore, the constant luciferin assumption is relaxed to some extent by varying the luciferin values as a function of the agent's position from a leader. This reflects the case where the first term in the luciferin update rule, given in (2.1), is zero (or $\rho = 1$).

Theorem 3.4 *Consider a single leader l such that the luciferin value of a follower is equal to that of the leader when it is co-located with the leader and decreases monotonically with distance from the leader, then there exists a sequence $\{f_1, f_2, \ldots, f_q\}$ in which the followers reach the leader and all the members of $N_l(t_0)$ are co-located with l, for all t_k, $k \geq K(q)$, where, $q = |F|$, $K(i) = \sum_{j=1}^{i} \left\lceil \frac{d_{f_j l}(t_{K(j-1)})}{\delta} \right\rceil$, $K(0) = 0$, and $f_j \in F$, $j = 1, \ldots, q$.*

Proof Consider a single leader l. Since $\ell_l > \ell_{f_j}$ for all $f_j \in N_l(t_0)$, all followers are influenced by the leader. Since the value of $\ell_{f_j}(t)$ is a function of distance between the leader l and the follower f_j at time t, it changes with time as f_j moves toward the leader. Consider three followers f_i, f_j, and f_k such that $d_{f_i l}(t) < d_{f_j l}(t) < d_{f_k l}(t)$ as shown in Fig. 3.12. Accordingly, we get

$$\ell_l > \ell_{f_i}(t) > \ell_{f_j}(t) > \ell_{f_k}(t) \tag{3.21}$$

Fig. 3.12 Positions of
followers f_i, f_j, and f_j at
time t and $t+1$, respectively

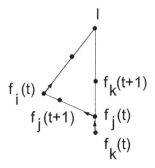

Since, $L_{f_i}(t) = \{l\}$, f_i moves toward l by a distance δ. Suppose f_j decides to move toward f_i and f_k decides to move toward l at time t, respectively. This gives

$$d_{f_i l}(t+1) < d_{f_k l}(t+1) < d_{f_j l}(t+1) \tag{3.22}$$

$$\Rightarrow \ell_{f_i}(t+1) > \ell_{f_k}(t+1) > \ell_{f_j}(t+1) \tag{3.23}$$

From (3.21) and (3.23), note that $\ell_{f_j}(t) > \ell_{f_k}(t)$, but $\ell_{f_j}(t+1) < \ell_{f_k}(t+1)$. Therefore, the followers cannot be sorted according to ascending order of their associated ℓ values as it was possible in Theorem 3.1. However,

$$d_{f_i l}(t) = d_{f_i l}(0) - \delta t \leq d_{f_j l}(0) - \delta t \leq d_{f_j l}(t) \tag{3.24}$$

Therefore, f_i always remains closer than f_j to the leader during its movements.
Now at time t_0, let

$$f_1 = \arg\{\min_{f_i \in F}\{d_{f_i l}(t_0)\}\} \tag{3.25}$$

From (3.25), $\ell_{f_1}(t_k) \geq \ell_{f_i}(t_k)$, for all $f_i \in F, i \neq 1$ and for all $t_k \geq t_0$. Therefore, f_1 makes deterministic movements toward l and reaches it in $t_{K(1)}$ iterations, where $K(1) = \left\lceil \frac{d_{f_1 l}(0)}{\delta} \right\rceil$. Therefore, for all $t_k, k \geq K(1)$, f_1 is co-located with l.
Next, identify the follower f_2 such that

$$f_2 = \arg\{\min_{f_i \in F,\ i \neq 1}\{d_{f_i l}(t_{K(1)})\}\} \tag{3.26}$$

Using analysis that is similar to that of the movements of follower f_2 in Theorem 3.1, it can be shown that for all $t_k, k \geq K(2)$, f_2 is co-located with l, where

$$K(2) = K(1) + \left\lceil \frac{d_{f_2 l}(t_{K(1)})}{\delta} \right\rceil \tag{3.27}$$

Now, suppose f_j is co-located with l, for all $t_k, k \geq K(j)$. Identify follower f_{j+1} such that

$$f_{j+1} = \arg\{\min_{f_i \in F - \{1,2,\dots,j\}} \{d_{f_i l}(t_{K(j)})\}\} \tag{3.28}$$

Let $d_{f_{j+1}l}(t_{K(i)})$ be the distance between f_{j+1} and l at time $t_{K(j)}$. Now, f_{j+1} takes $\lceil d_{f_{j+1}l}(t_{K(j)}) \rceil$ steps to reach l. This implies,

$$
\begin{aligned}
K(j+1) &= K(j) + \left\lceil \frac{d_{f_{j+1}l}(t_{K(j)})}{\delta} \right\rceil \\
&= \sum_{i=1}^{j} \left\lceil \frac{d_{f_i l}(t_{K(i-1)})}{\delta} \right\rceil + \left\lceil \frac{d_{f_{j+1}l}(t_{K(j)})}{\delta} \right\rceil \\
&= \sum_{i=1}^{j+1} \left\lceil \frac{d_{f_i l}(t_{K(i-1)})}{\delta} \right\rceil
\end{aligned}
$$

Therefore, by induction, it can be shown that there exists a sequence $\{f_1, f_2, \dots, f_q\}$ in which the followers reach the leader and all the members of $N_l(t_0)$ will be co-located with l, for all $t_k, k \geq K(q)$. $\square$

Next, two leaders l_1 and l_2 and a set of followers F where the luciferin value of each follower is assumed to be a function of its distance from the two leaders are considered. The analysis is restricted to a one-dimensional case. In particular, the following function is used to model the luciferin values of the followers:

$$\ell(x) = a_1 e^{-b_1|x-x_1|} + a_2 e^{-b_2|x-x_2|} \tag{3.29}$$

where, a_1, b_1, a_2, and b_2 are positive constants.

Theorem 3.5 *Consider two leaders l_1 and l_2 located at positions x_1 and x_2, respectively, on a real axis such that $0 \leq x_1 \leq x_2 - r_s$ and a set of followers F such that $x_1 < x_{f_i} < x_2$ for all $f_i \in F$. If the luciferin value of each follower varies according to (3.29), then the line segment $\overline{x_1 x_2}$ can be partitioned into three regions $R_a = [x_1, x_i), R_b = [x_i, x_i + r_s]$, and $R_c = (x_i + r_s, x_2]$ where,*

$$x_i = \frac{b_1 x_1 + b_2 x_2 - \ln\left[\frac{a_2(e^{b_2 r_s} - 1)}{a_1(1 - e^{-b_1 r_s})}\right]}{b_1 + b_2} \tag{3.30}$$

and all the followers in regions R_a and R_c converge to the leaders l_1 and l_2, respectively, in finite time and all the followers in region R_b asymptotically converge to one of the leaders.

Proof Consider two leaders l_1 and l_2 located at x_1 and x_2, respectively, on the real axis such that $0 \leq x_1 \leq x_2 - r_s$.

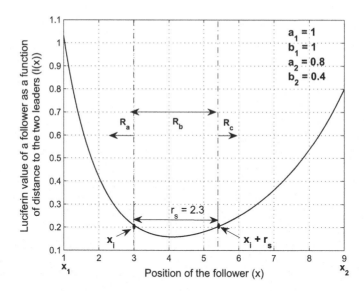

Fig. 3.13 Plot of $\ell(x)$ (3.31) for $x \in [x_1, x_2]$ when the values given in Table 3.3 are used for $a_1, b_1, a_2,$ and b_2

Table 3.3 Values chosen for the constants in the luciferin function to generate Fig. 3.13

a_1	b_1	a_2	b_2	x_1	x_2
1	1	0.8	0.4	1	9

Assuming that the luciferin value of each follower is given by (3.29) and given that each follower f_i lies on the line segment $\overline{x_a x_b}$, (3.29) that holds for the region $[x_1, x_2]$ can be rewritten as below:

$$\ell(x) = a_1 e^{-b_1(x-x_1)} + a_2 e^{-b_2(x_2-x)} \tag{3.31}$$

where, $x_1 \leq x \leq x_2$. An example plot of the above function is shown in Fig. 3.13. The values chosen for the various constants are shown in Table 3.3. The luciferin profile $\ell(x)$ (3.31) is a convex function and has a property that its value monotonically decreases as we move from x_{l_1} toward l_2 and reaches a unique minimum at some point x^* in the interval $[x_1, x_2]$. Thereafter, it increases monotonically until x_{l_2}.

x^* can be solved as follows:

Using the first order sufficient conditions for a minimum,

$$\frac{d\ell(x)}{dx}\Big|_{x=x^*} = -a_1 b_1 e^{-b_1(x^*-x_1)} + a_2 b_2 e^{-b_2(x_2-x^*)} = 0$$

$$e^{b_1(x^*-x_1)-b_2(x_2-x^*)} = \frac{a_1 b_1}{a_2 b_2}$$

Applying natural logarithm on both sides of the above equation,

$$(b_1 + b_2)x^* - (b_1x_1 + b_2x_2) = \ln\left[\frac{a_1b_1}{a_2b_2}\right]$$

$$\Rightarrow x^* = \frac{b_1x_1 + b_2x_2 - \ln\left[\frac{a_1b_1}{a_2b_2}\right]}{b_1 + b_2} \tag{3.32}$$

Note that any line parallel to the x-axis intersects the luciferin profile at two points $(x_i, \ell(x_i))$ and $(x_j, \ell(x_j))$ such that $\ell(x_i) = \ell(x_j)$.

Now, consider x_i and $x_i + r_s$ such that

$$\ell(x_i) = \ell(x_i + r_s). \tag{3.33}$$

x_i can be solved for in the following way. Using (3.29) and (3.33),

$$a_1e^{-b_1(x_i-x_1)} + a_2e^{-b_2(x_2-x_i)} = a_1e^{-b_1(x_i+r_s-x_1)} + a_2e^{-b_2(x_2-x_i-r_s)}$$
$$= a_1e^{-b_1(x_i-x_1)}e^{-b_1r_s} + a_2e^{-b_2(x_2-x_i)}e^{b_2r_s}$$
$$\Rightarrow a_1e^{-b_1(x_i-x_1)}(1 - e^{-b_1r_s}) = a_2e^{-b_2(x_2-x_i)}(e^{b_2r_s} - 1)$$

Taking exponential terms to one side and rewriting the above equation,

$$e^{-b_1(x_i-x_1)+b_2(x_2-x_i)} = \frac{a_2(e^{b_2r_s} - 1)}{a_1(1 - e^{-b_1r_s})} \tag{3.34}$$

Applying natural logarithm on both sides of the above equation,

$$-(b_1 + b_2)x_i + b_1x_1 + b_2x_2 = \ln\left[\frac{a_2(e^{b_2r_s} - 1)}{a_1(1 - e^{-b_1r_s})}\right]$$

$$x_i = \frac{b_1x_1 + b_2x_2 - \ln\left[\frac{a_2(e^{b_2r_s}-1)}{a_1(1-e^{-b_1r_s})}\right]}{b_1 + b_2} \tag{3.35}$$

For the example in Fig. 3.13, $x_i = 3$ for $r_s = 2.3$.

Now consider followers f_a, f_b, and f_c at time t such that $x_{f_a}(t) \in [x_1, x_i), x_{f_b}(t) \in [x_i, x_i + r_s]$, and $x_{f_c}(t) \in (x_i + r_s, x_2]$. From (3.33) and using the convex property of $\ell(x)$, we have

$$x^* \in [x_i, x_i + r_s] \tag{3.36}$$
$$\ell_{f_a}(t) > \ell_{f_b}(t) \text{ for all } f_a \in R_a, \text{ for all } f_b \in R_b \tag{3.37}$$

From (3.36) and (3.37), f_b does not influence the movement of f_a. Moreover, since $d_{f_af_c}(t) > r_s$, for all $f_c \in R_c$, f_a cannot sense the presence of, and its movements are not influenced by, f_c. These two conditions, prevent f_a to move toward the leader l_2. However, it is provided with an uphill direction toward the leader l_1. Using the analysis from Theorem 3.4, it can be proved that f_a reaches l_1 in finite number of time steps. Similar analysis can be used for the follower f_c to prove that it converges to the leader l_2 in finite number of time steps.

However, for all $f_b \in R_b$, f_b is influenced by followers in R_a and followers in R_c as well. Hence, f_c has an uphill direction toward both the leaders until a time t_k when it reaches either x_i or $x_i + r_s$. Using the result from Theorem 3.3, it can be shown that f_b reaches one of the leaders asymptotically.   □

Note that though Theorem 3.5 was proved for a composite exponential function, it can be extended to include most monotonically decaying functions which on combining produce a convex profile between the two leaders.

3.5 Simulations on Simplified GSO Model

A set of four experiments are conducted in order to verify the theoretical results arrived at in the previous section. In all the experiments, each agent is associated with a constant luciferin level and a constant local-decision domain ($r_d^i = 1.5$, for all $i = 1, \ldots n$). The first three experiments are designed to explicitly represent the assumptions of isolated, non-isolated and non-overlapping, and overlapping leader-neighborhoods that are made in the first three theorems, respectively. In the last experiment, a more general placement of agents is considered and identification of leaders and followers based on whether they move or remain stationary for all time is attempted, and the various leader-neighborhoods are characterized based on the inter-leader distances.

3.5.1 Simulation Experiment 1: Isolated Leader

A set of 9 agents is randomly deployed in a circle of radius 1.5 units and centered at the leader's location $(0, 0)$. The luciferin value of the leader is chosen as $\ell_l = 10$. The emergence of the agent movements is shown in Fig. 3.14a. Agents are ranked according to their luciferin levels. For instance, Agent 1 has the highest luciferin value among all the followers. Table 3.4 shows the luciferin levels of the agents, distance of agent j when agent $j - 1$ reaches the leader $d_{jl}(t_{K(j-1)})$ at time $t_{K(j-1)}$, and the number of iterations taken by agent j to reach the leader ($K(j)$). From Theorem 3.1, $K(9) = \sum_{j=1}^{9} \left\lceil \frac{d_{jl}(t_{K(j-1)})}{s} \right\rceil$.

Table 3.4 Luciferin levels of the agents, distance of agent j when agent $j - 1$ reaches the leader $d_{jl}(t_{K(j-1)})$ at time $t_{K(j-1)}$, and the number of iterations taken by agent j to reach the leader $(K(j))$

Agent	1	2	3	4	5	6	7	8	9
ℓ_i	8.12	6.98	6.63	6.59	6.46	6.10	5.77	4.93	4.29
$d_{jl}(t_{K(j-1)})$	0.539	0.224	0.023	0	0.024	0.116	0.045	0.199	0.097
$t_{K(j)}$	18	26	27	27	28	32	35	41	45

Using the values from Table 3.4, we get

$$K(9) = \left\lceil \frac{0.5385}{0.03} \right\rceil + \left\lceil \frac{0.2237}{0.03} \right\rceil + \left\lceil \frac{0.0234}{0.03} \right\rceil + \left\lceil \frac{0.0241}{0.03} \right\rceil + \left\lceil \frac{0.1158}{0.03} \right\rceil$$
$$+ \left\lceil \frac{0.0449}{0.03} \right\rceil + \left\lceil \frac{0.1985}{0.03} \right\rceil + \left\lceil \frac{0.0986}{0.03} \right\rceil$$
$$= 45$$

Note that the above result, computed using Theorem 3.1, coincides with the simulation result given in Table 3.4.

3.5.2 Simulation Experiment 2: Non-isolated and Non-overlapping Leader Neighborhoods

Two leaders are located at $(-2, 0)$ and $(2, 0)$, with $\ell_{l_1} = 100$ and $\ell_{l_2} = 50$, respectively. Note that the above placement of leaders leads to non-isolated and non-overlapping leader-neighborhoods in the sense defined in the previous section. A set of 98 agents is randomly deployed such that all agents located inside the circle of radius 1.5 units and centered at $(2, 0)$ have a luciferin level less than 50 and the remaining agents have a luciferin level less than 100. This ensures that L_2 remains stationary for all time since any agent with a luciferin level more than 50 is outside its sensing range of 1.5 units. The emergence of agent-movements is shown in Fig. 3.14b. All agents with $\ell > 50$ reach the leader L_1 and most of the agents that are located inside the neighborhood of L_2 converge to L_2. However, note that a few agents leave the L_2-neighborhood, enter the L_1-neighborhood, and get co-located at L_1. This is in perfect agreement with the observation made in the proof of Theorem 3.2 that an agent can leave the neighborhood of L_2 and once it enters the L_1-neighborhood, it remains in the same neighborhood, and eventually converges to L_1.

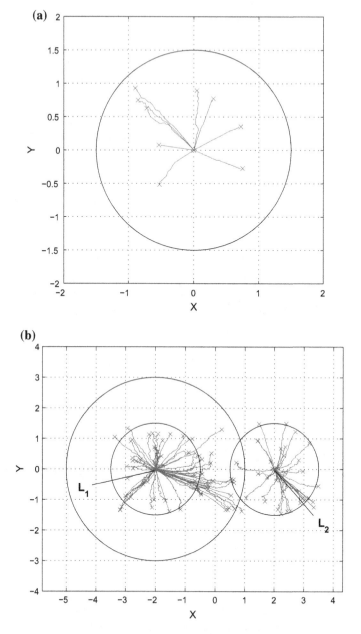

Fig. 3.14 a Agent movements in an isolated leader's neighborhood. **b** Emergence of agent movements in the case of non-isolated and non-overlapping leader-neighborhoods

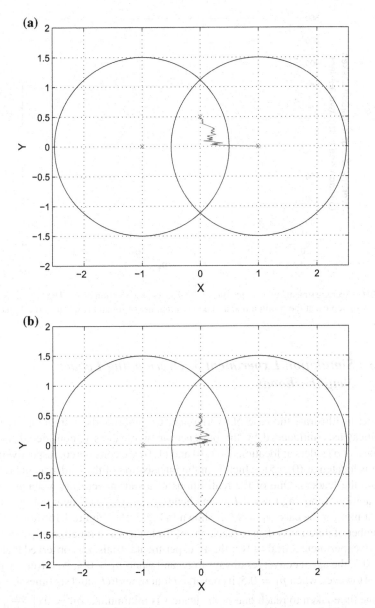

Fig. 3.15 Overlapping leader-neighborhoods. Agent movements when **a** $p_1 = 0.45$. **b** $p_1 = 0.6$

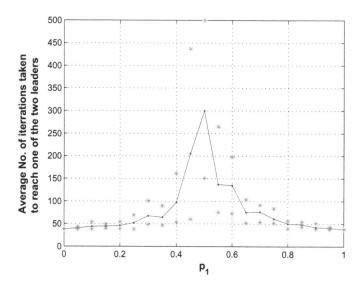

Fig. 3.16 Average time taken to reach one of the leaders as a function of p_1. The '*' marks at each value of p_1 represent the minimum and maximum number of iterations taken to reach one of the leaders

3.5.3 Simulation Experiment 3: Overlapping Leader Neighborhoods

In order to illustrate the case of overlapping neighborhoods, a single agent that is simultaneously influenced by the presence of two leaders is considered. For this purpose, two leaders at locations $(-1, 0)$ and $(1, 0)$ are considered, respectively. The agent is located at $(0, 0.5)$, which is in the intersection of the leader-neighborhoods of both the leaders. One of the realizations of agent movements when $p_1 = 0.45$ (probability of moving toward L_1), and hence $p_2 = 0.55$, is shown in Fig. 3.15a. A similar plot for the case $p_1 = 0.6$ is shown in Fig. 3.15b. Figure 3.16 shows the plot of number of iterations taken to reach one of the leaders as a function of p_1, averaged over 10 experimental trials. Over the 10 experimental trials, it is observed that when $p_1 < 0.5$, the agent converges always to L_2 and when $p_1 > 0.5$, it converges always to L_1. However, when $p_1 = 0.5$, it converged four times to L_1 and six times to L_2. The average time taken to reach one of the leaders is minimum at $p_1 = 0$ ($\left\lceil \frac{\sqrt{0.5^2+1}}{0.03} \right\rceil = 38$), increases with increase in p_1, reaches a maximum at $p_1 = 0.5$, and then decreases reaching a minimum again at $p_1 = 1$. The result in Fig. 3.16 signifies the fact that the randomness and the asymptotic behavior of the agent are more prominent in the vicinity of $p_1 = 0.5$. This is evident from the large variability and relatively more convergence time in this region. The agent movements become more deterministic as p_1 tends away from 0.5.

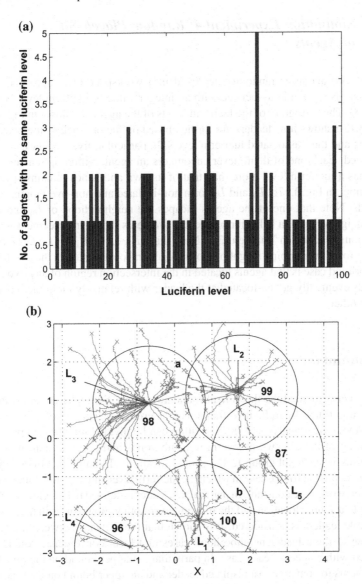

Fig. 3.17 a Histogram of luciferin levels of the agents. **b** Formation of leaders and followers based on the placement of agents and their associated luciferin levels. The numbers represent the luciferin levels associated with the leaders

3.5.4   Simulation Experiment 4: Random Placement
of Agents

A set of 100 agents is randomly deployed in a workspace of $(-5, 5) \times (-5, 5)$. Each agent is randomly associated with an integer value of luciferin level between 1 and 100. The histogram of the luciferin levels of the agents is shown in Fig. 3.17a. Figure 3.17b shows how leaders are formed based on factors such as the placement of agents and their associated luciferin levels. In particular, five leaders $L_1, \ldots, L_5$ are formed. Each one of them never encounters an agent, within its sensing range, which has a luciferin level more than that of its own and hence remains stationary for all time. In Fig. 3.17b, L_1 and L_3 form non-isolated and non-overlapping neighborhoods. Note that one of the agents escapes the neighborhood of L_3 and enters the L_1-neighborhood, a behavior of agent movements that is in agreement with the conclusions reached in Theorem 3.2. Note that (L_1, L_4), (L_1, L_5), (L_2, L_3), and (L_2, L_5) form overlapping neighborhoods. A general observation in the overlapping neighborhood case is that agents located in the intersection region of any two neighborhoods eventually get co-located at the leader with relatively more luciferin level than the other.

3.6   Summary

This chapter presented the theoretical foundations for GSO under certain restricted set of assumptions. A careful investigation of the group-level phases of GSO revealed that the theoretical analysis of the algorithm can be divided into two subproblems: (1) splitting of the glowworm swarm into subgroups and (2) local convergence of glowworms in each subgroup to the source locations. Initially, local convergence results were proved for the case in which the luciferin level and decision domain range of the glowworms are constant. In particular, a leader is defined to be a glowworm that is closer to a source location and the movements of other glowworms in the presence of multiple leaders is considered. It is shown that the distance between the leaders gives rise to three leader neighborhood cases and local convergence results were presented with respect to each case. In particular, an upper bound on the time taken by the agents to converge to an isolated leader and an upper bound on the time taken by the agents to converge to one of the leaders with non-isolated and non-overlapping neighborhoods were determined. It was shown that glowworms under the influence of multiple leaders with overlapping neighborhoods asymptotically converge to one of the leaders. Later, the constant luciferin assumption was relaxed to some extent and similar results were proved. Finally, the theoretical results were illustrated using simulations. In the next chapter, numerical simulation experiments are used to test GSO on a number of benchmark functions to validate its applicability for multimodal function optimization.

3.7 Thought and Computer Exercises

Exercise 3.1 Consider the multimodal function given in (3.1) in a search space of $[-5, 5] \times [-5, 5]$. Set $Q = 10$. Set equal slope profiles by using $b_i = 3$. Instantiate values for a_i, x_i, and y_i ($i = 1, 2, \ldots 10$) using (3.2)–(3.4). Run the GSO code and plot the emergence of agent movements (similar to the graphs in Fig. 3.2). Overlay a Voronoi partition of the peak locations on the previous plot (Hint: You can use the MATLAB function 'voronoi(X,Y)' for this purpose). How many peaks are captured? Comment on the splitting pattern of the glowworm swarm. What is the fraction of glowworms whose initial location lie in one partition but their final locations lie in a different partition?

Exercise 3.2 Repeat the experiment in the previous exercise 10 times using the same initial placement of the glowworms. Report the mean and standard error of mean (SEM) values for (1) the number of peaks captured and (2) the fraction of glowworms that ended up in different partitions. Use these results to comment on how the algorithm handles profiles with closely spaced peaks, distant peaks, and peaks on the edges of the workspace.

Exercise 3.3 Select one multimodal function instantiation from Exercise 3.1 and run the GSO code for 30 trials using different uniformly distributed random initial placement of glowworms. Record the number of peaks captured in each case. Compute the mean and SEM values and use these results to comment on the sensitivity to initial placements of glowworms.

Exercise 3.4 Consider the multimodal function (3.1) again. Set $Q = 2$, $a_1 = 4$, $a_2 = 2$, $(x_1, y_1) = (-3, 0)$, and $(x_2, y_2) = (3, 0)$. Set unequal slope profiles by using $b_1 = 0.9$ and $b_2 = 0.2$. Run the GSO code and plot the emergence of agent movements. Overlay an equi-valued contour on the previous plot (Hint: Use the MATLAB function 'contour(X,Y,Z)' for this purpose). Comment on the splitting pattern of the glowworm swarm in this new regime. What causes the change in the splitting behavior (if any) from the regime used in the Exercise 3.1?

Exercise 3.5 Consider the radially-symmetric function profile $J(x, y) = -(x^2 + y^2)$ with initial placements of two glowworms G_a and G_b at $(-5, -5)$ and $(12, 12)$, respectively. Set the GSO parameters $\gamma = 1$, $\rho = 1$, $n_t = 1$, and $s = 0.3$. Further, assume that the sensing range r_s and the initial neighborhood range $r_d^i(0)$ ($i = 1, 2$) are equal to the distance between initial placements of the two glowworms. Now, answer the following questions analytically (without resorting to computer simulations).

3.5.1 Show that the given settings lead to deterministic movements of the two glowworms (Hint: Rewrite the GSO equations (2.1)–(2.4) using the information provided and analyze how each glowworm updates its luciferin, movement, and neighborhood range). Show that the glowworm movements are collinear. What is the convex hull formed by the positions of the two glowworms? Does the peak $(0, 0)$ lie in it?

3.5.2 Which of the two glowworms makes the first movement? Explain. Determine the condition when the other glowworm make its first movement? (Hint: Draw circular equi-valued contours at appropriate places and analyze the movement decisions) How many steps elapse before this condition is met? Describe the pattern of movements of the glowworm pair after this condition is met.

3.5.3 Show that the glowworm G_b converges to $(0, 0)$ in finite time. Does G_a converge to $(0, 0)$ too? If not, how does its convergence behavior look like? What happens to their convergence behavior when the step size s is changed to 0.5?

Exercise 3.6 Consider the same settings used in Exercise 3.5. Now, shift the initial placements of G_a and G_b to $(-5, -8)$ and $(12, 9)$, respectively. Does $(0, 0)$ lie in the convex hull formed by the new initial placements of the glowworms? Determine the convergence behavior of the glowworm pair. Add a new glowworm G_c at $(-5, 5)$. Determine the convex hull of the initial placements of G_a, G_b, and G_c and check if $(0, 0)$ lies in it. How does the convergence behavior of the glowworms change in this new regime? Next, shift the initial placement of G_c to $(5, -5)$. Repeat the analysis for this new initial placement of the glowworms.

Exercise 3.7 Support your arguments in Exercises 3.5 and 3.6 through computer simulations. Use your observations so far to comment about the relationship between the initial placements of glowworms and their convergence behavior.

Chapter 4
Multimodal Function Optimization

In the previous chapter, the theoretical formulation of a simplified model of GSO was presented. In this chapter, numerical simulation results to evaluate the efficacy of GSO in capturing multiple optima of multimodal optimization problems are presented. Experiments are restricted to maximization problems. However, the algorithm can be easily modified and used for minimization problems. For this purpose, several benchmark multimodal functions are considered that pose the following cases:

- unequal peaks
- equal peaks
- peaks of concentric circles
- peaks surrounded by regions with step-discontinuities (non-differentiable objective functions)
- peaks comprising plateaus of equal heights
- peaks located at irregular intervals
- change in landscape features with change in scale
- non-separability involving interdependence of objective function variables

The parameter selection problem is addressed by conducting experiments to show that only two parameters need to be selected by the user. Next, simulation experiments are used to compare GSO with Niche-PSO, a PSO variant that is designed for the simultaneous computation of multiple optima. Finally, algorithmic behavior is examined in the presence of noise.

4.1 Multimodal Test Functions

In the previous chapters, J_1 (2.24) and J_2 (3.1) functions have been used to illustrate the basic working mechanism of GSO. In this chapter, GSO is tested on the following benchmark multimodal functions: Rastrigin's, Circles, Staircase, Plateaus,

© Springer International Publishing AG 2017
K.N. Kaipa and D. Ghose, *Glowworm Swarm Optimization,*
Studies in Computational Intelligence 698, DOI 10.1007/978-3-319-51595-3_4

Equal-peaks-A, Equal-peaks-B, Equal-peaks-C, Himmelblau's, Schwefel's, Griewangk's, Ackley's, Langerman's, Michalewicz's, Drop wave, Shubert's, and the fifth function of De Jong. Multiple sets of test functions are chosen for different purposes: tests in two dimensions, tests to evaluate GSO parameters on algorithmic performance, tests in higher dimensions, and experiments based comparisons between GSO and a state-of-the-art PSO variant used to capture multiple optima.

Let $X = [x_1, x_2, \ldots, x_m] \in \Re^m$ be a position vector in the m-dimensional search space with $m \geq 2$. Using this notation, the multimodal functions are defined as follows.

Rastrigin's function:

$$J_3(X) = \sum_{i=1}^{m}(10 + x_i^2 - 10\cos(2\pi x_i)) \tag{4.1}$$

The Rastrigin's function (Fig. 4.1) was first proposed by Rastrigin [207] as a 2-dimensional function and has been generalized by Mühlenbein et al. in [157]. This function presents a fairly difficult problem due to its large search space and its large number of local minima and maxima. For instance, while the Peaks function (J_1), given in (2.24), consists of only three peaks, the Rastrigin's (J_3) function consists of as many as 100 peaks in the considered range. The Rastrigin's function has been used as a benchmark function to test several algorithms designed to solve global and multimodal function optimization problems [159, 208].

Circles function:

$$J_4(x_1, x_2) = (x_1^2 + x_2^2)^{0.25}((\sin^2(50(x_1^2 + x_2^2)^{0.1})) + 1.0) \tag{4.2}$$

The Circles function [159] contains multiple concentric circles as the regions of local maxima (Fig. 4.2). Unlike J_1, J_2, and J_3 functions where each peak is a single point, the circular lines of local peaks present a infinite-peaks case.

Staircase function:

$$J_5(x_1, x_2) = 25 - \lfloor x_1 \rfloor - \lfloor x_2 \rfloor \tag{4.3}$$

The Staircase function [49] contains a series of stairs as shown in Fig. 4.3. In (4.3), $\lfloor x \rfloor$ is the floor function and is defined as the nearest integer lesser than x. Peaks are flat regions that are surrounded by step-discontinuities. The Staircase function presents a case where the function value increases in discrete steps as we move, from one stair to the next, towards the tallest stair (global peak region).

Plateaus function:

$$J_6(x_1, x_2) = \text{sign}(\cos(x_1) + \cos(x_2)) \tag{4.4}$$

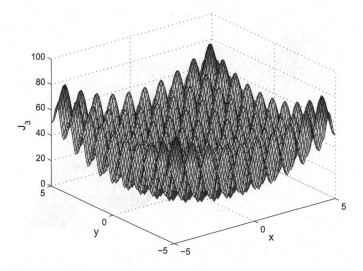

Fig. 4.1 Rastrigin's function in two dimensions

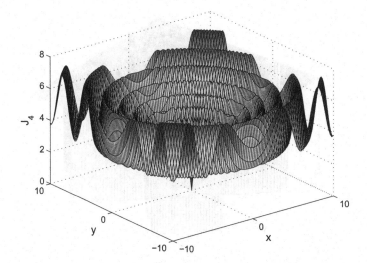

Fig. 4.2 Circles function in two dimensions

The Plateaus function contains multiple plateaus (Fig. 4.4). Even though the Plateaus function is similar to the Staircase function in terms of the nature of peaks (both have flat regions as peaks), the plateaus in the J_6 function have equal objective function values, thereby providing no uphill direction.

Equal-peaks-A function:

$$J_7(X) = \sum_{i=1}^{m} \cos^2(x_i) \tag{4.5}$$

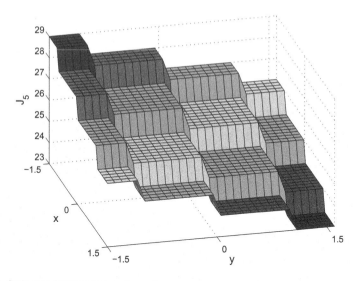

Fig. 4.3 Stair-case function in two dimensions

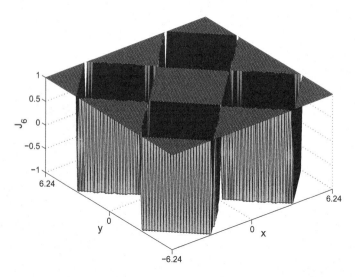

Fig. 4.4 Plateaus function in two dimensions

$J_7(X)$ is called the Equal-peaks function as all local maxima have equal function values (Fig. 4.5). We use $J_7(X)$ in the experiments used to evaluate the effect of n and r_s on algorithmic performance.

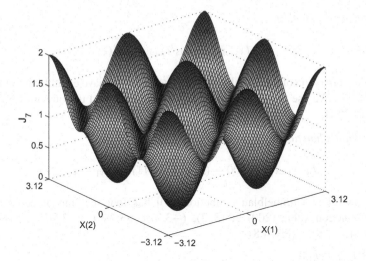

Fig. 4.5 Equal-peaks-A function in two dimensions

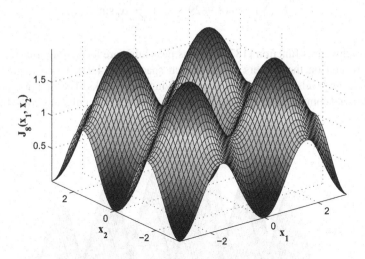

Fig. 4.6 Equal-peaks-B function in two dimensions

Equal-peaks-B function:

$$J_8(X) = \sum_{i=1}^{m} \sin^2(x_i) \tag{4.6}$$

The shape of the function profile of the Equal-peaks-B function (Fig. 4.6) is similar to that of the Equal-peaks-A function. We use the function $J_7(X)$ for tests on GSO in higher dimensional search spaces.

Equal-peaks-C function:

$$J_9(x_1, x_2) = \cos^2(x_1) + \sin^2(x_2) \qquad (4.7)$$

All local maxima of the Equal-peaks-C function [168] have equal function values (Fig. 4.7). We use the $J_9(x_1, x_2)$ in the experiments used to compare GSO with Niche-PSO.

Himmelblau's function:

$$J_{10}(x_1, x_2) = 200 - (x_1^2 + x_2^2 - 11)^2 - (x_1 + x_2^2 - 7)^2$$

The modified Himmelblau's function [20] has four maxima of equal objective function value (200) at $(3, 2)$, $(-3.779, -3.283)$, $(-2.805, 3.131)$, and $(3.584, -1.848)$ (Fig. 4.8).

Schwefel's function:

$$J_{11}(X) = 418.9829 \times m + \sum_{i=1}^{m} x_i \sin(\sqrt{|x_i|}) \qquad (4.8)$$

The Schwefel's function [187] contains peaks that are located interior to, on the edges, and on the corners of the search space (Fig. 4.9a), thereby, presenting increasing levels of complexity in identifying these peaks. The function presents additional complexity in that the peaks are not located at regular intervals; both the

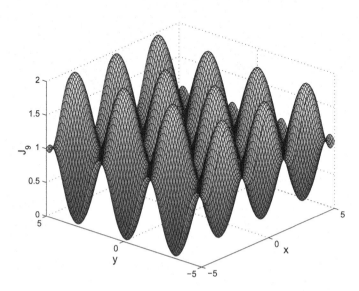

Fig. 4.7 Equal-peaks-C function in two dimensions

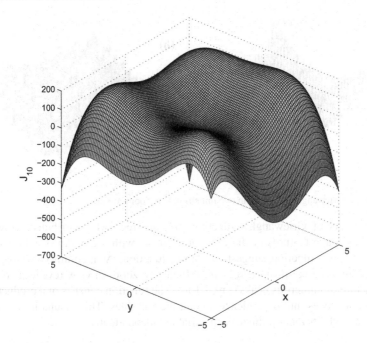

Fig. 4.8 Himmelblau's function in two dimensions

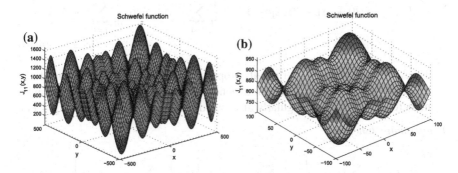

Fig. 4.9 a Schwefel's function in two dimensions. **b** Zoom-in view

inter-peak distance and peak-heights (function value at the peak location) increase with distance from the origin along each coordinate axis (Fig. 4.9b).

Griewangk's function:

$$J_{12}(X) = 1 + \sum_{i=1}^{m} \frac{x_i^2}{4000} - \prod_{i=1}^{m} \cos\left(\frac{x_i}{\sqrt{i}}\right) \qquad (4.9)$$

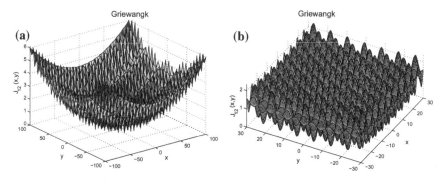

Fig. 4.10 **a** Griewangk's function in two dimensions. **b** Zoom-in view

The peaks of Griewangk's function [67] are regularly distributed. However, the landscape features of the function changes with scale. An overview of the Griewangk's function suggests a convex function. At medium-view scale, the function has local-peaks (Fig. 4.10a). Finally, a zoom-in view reveals a complex structure of numerous peaks (Fig. 4.10b). The function consists of a product term that introduces interdependence among the variables. This results in the failure of the methods that optimize each variable independently.

Ackley's function:

$$J_{13}(X) = 20 + \exp -20 \exp(-0.2\sqrt{(\frac{1}{m}\sum_{i=1}^{m}x_i^2)}) - \exp(\frac{1}{m}\sum_{i=1}^{m}\cos(2\pi x_i))$$

(4.10)

The Ackley's function was originally proposed by Ackley [1] and was generalized by Bäck [9]. The function has an exponential term that covers its surface with numerous peaks that are regularly distributed across the search space (Fig. 4.11).

Langerman's function:

$$J_{14}(X) = \sum_{i=1}^{p} c_i \exp\left[-\frac{1}{\pi}\sum_{i=1}^{m}(x_j - a_{ij})^2\right] \cos\left[\pi\sum_{i=1}^{m}(x_j - a_{ij})^2\right]$$

(4.11)

where, $(c_i, i = 1, 2, \ldots, p)$, $(a_{ij}, j = 1, 2, \ldots, m)$ are constant numbers fixed in advance. The symmetry advantages offered by symmetrical multimodal functions might simplify optimization for certain algorithms. However, the Langerman's function [192] is non-symmetrical and its local optima are randomly distributed. Figure 4.12 shows an instance of Langerman's function that is generated when the following values are used for the function parameters:

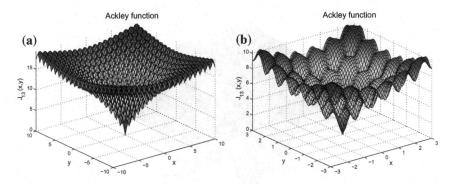

Fig. 4.11 a Ackley's function in two dimensions. **b** Zoom-in view

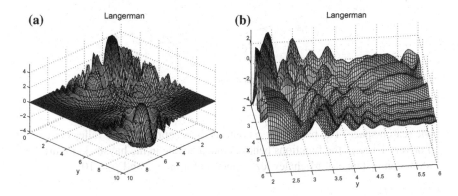

Fig. 4.12 a Langerman's function in two dimensions. **b** Zoom-in view

$$p = 5$$
$$m = 2$$
$$C = \begin{bmatrix} 1 & 2 & 5 & 2 & 3 \end{bmatrix}^T$$
$$A = \begin{bmatrix} 3 & 5 & 2 & 1 & 7 \\ 5 & 2 & 1 & 4 & 9 \end{bmatrix}^T$$

Michalewicz's function:

$$J_{15}(X) = -\sum_{i=1}^{m} \sin(x_i) \left[\sin\left(\frac{i x_i^2}{\pi}\right) \right]^{2p} \qquad (4.12)$$

The landscape of Michalewicz's function [9] in two dimensions is composed of steep valleys and approximately flat regions where the function value increases with a very small slope as we move toward the peaks on the edges and corners of

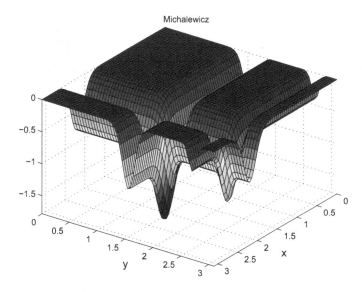

Fig. 4.13 Michalewicz's function in two dimensions

the search space. The parameter p controls the steepness of the valleys or edges. For very large p, the functions behaves like a needle in the haystack. Figure 4.13 shows an instance of Michalewicz's function for $m = 2$ and $p = 10$.

Drop wave function:

$$J_{16}(x_1, x_2) = -\frac{1 + \cos(12\sqrt{(x_1^2 + x_2^2)})}{\frac{1}{2}(x_1^2 + x_2^2) + 2} \tag{4.13}$$

A landscape overview of the Drop wave function is similar to the Circles function (Fig. 4.14a). However, a closeup view will reveal numerous local peaks that lie on concentric circles centered about the origin of the search space (Fig. 4.14b).

Shubert's function:

$$J_{17}(x_1, x_2) = -\sum_{i=1}^{5} i \cos((i + 1)x_1 + 1) \sum_{i=1}^{5} i \cos((i + 1)x_2 + 1) \tag{4.14}$$

The Shubert's function [150] consists of an interesting pattern of peaks. Whereas peaks are distributed across the entire search space, they can be classified into two types. The first set of peaks form a '+' pattern, which is aligned along the two orthogonal coordinate axes centered around the origin (Fig. 4.15a). The second set of peaks, with relatively smaller peak-heights are distributed regularly throughout the rest of the search space (Fig. 4.15b).

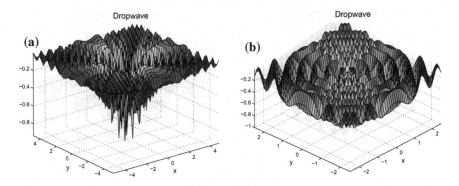

Fig. 4.14 a Dropwave function in two dimensions. **b** Zoom-in view

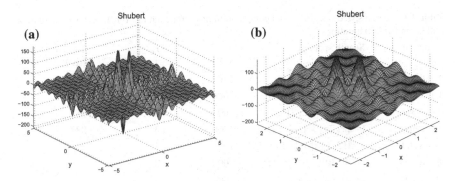

Fig. 4.15 a Shubert's function in two dimensions. **b** Zoom-in view

Fifth function of De Jong:

$$J_{18}(x_1, x_2) = -\left\{0.002 + \sum_{i=-2}^{2} \sum_{j=-2}^{2} \left[5(i+2) + j + 3 + (x_1 - 16j)^6 + (x_2 - 16i)^6\right]^{-1}\right\}^{-1}$$

The landscape of the Fifth function of De Jong [86] is composed of steep and equally spaced valleys, with the junctions between the valleys giving rise to peaks arranged on a two-dimensional grid (Fig. 4.16).

The search space size considered for each function, the values of parameters n and r_s, and the corresponding computation times are given in Table 4.1. Note that the computation times depend on several factors such as the workspace-size, nature of peaks, number of peaks, and number of agents used.

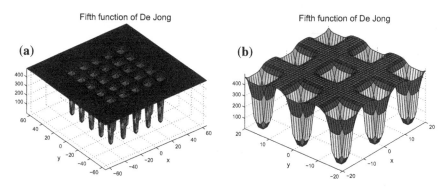

Fig. 4.16 a Fifth function of De Jong in two dimensions. **b** Zoom-in view

Table 4.1 Search space size, n, r_s, and computation times for each test function. Each simulation trial is run for 250 iterations

Function	Search space size	n	r_s	Total time (sec)[†]
Rastrigin's	$[-5, 5] \times [-5, 5]$	1500	2	1534
Circle's	$[-10, 10] \times [-10, 10]$	1000	4	584
Staircase	$[-2, 2] \times [-2, 2]$	1000	0.75	557
Plateaus	$[-2\pi, 2\pi] \times [-2\pi, 2\pi]$	1000	3	883
Schwefel's	$[-512, 512] \times [-512, 512]$	1000	3	
Griewangk's	$[-30, 30] \times [-2\pi, 2\pi]$	1000	3	
Ackley's	$[-10, 10] \times [-2\pi, 2\pi]$	1000	3	
Langerman's	$[0, 10] \times [0, 10]$	1000	3	883
Michalewicz's	$[0, \pi] \times [0, \pi]$	1000	3	
Drop wave	$[-5.12, 5.12] \times [-5.12, 5.12]$	1000	3	
Shubert's	$[-65.536, 65.536] \times [-65.536, 65.536]$	1000	3	
Fifth function of De Jong	$[-2\pi, 2\pi] \times [-2\pi, 2\pi]$	1000	3	

[†]The algorithm is coded using Matlab 7.0 and all simulations are implemented on a 2 GB RAM, 3 GHz Intel P4 machine with a Windows XP operating system

4.2 Performance Measures

To evaluate the performance of GSO, we use the fraction of peaks captured and the average minimum distance to the peak locations. We consider a glowworm to be located at a peak when its position is within a small ϵ-distance from the peak and we assume that a peak is captured when at least three glowworms are located at the peak. The value $\epsilon = 0.05$ is chosen for all experiments. The average minimum distance to the peak locations $d_{min,av}$ is given by:

$$d_{min,av} = \frac{1}{n} \sum_{i=1}^{n} \min_{\{1,...,Q\}} \{d_{i1}, ..., d_{iQ}\} \tag{4.15}$$

where, $d_{ij} = \|x_i - p_j\|$, $i = 1, ..., n$, $j = 1, ..., Q$, x_i and p_j are the locations of glowworm i and peak j, respectively, and Q is the number of peak locations.

4.3 Simulation Experiment Results

4.3.1 Tests on the Rastrigin's Function

A set of 1500 agents is deployed in a region of 10×10 units centered at $(0, 0)$ that contains a total number of 100 peaks (Fig. 4.1). Figure 4.17a and b show the

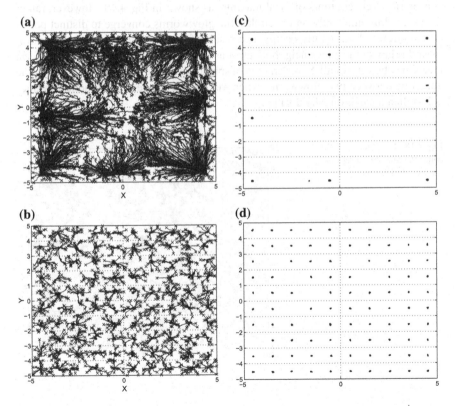

Fig. 4.17 Solution of the Rastrigin's function. **a** When the local-decision domain range r_d^i is kept constant. **b** When the local-decision domain range r_d^i is made adaptive Co-location of groups of glowworms at various peaks. **c** In constant decision domain case. **d** In the adaptive decision domain case

trajectories (250 iterations) followed by the glowworms from initial random locations
to one of the peaks, for the constant and adaptive neighborhood cases, respectively.
Figure 4.17c and d show the final co-location of groups of glowworms at various
peaks for the above two cases, respectively. Note that 92% of the peaks are captured
in the variable decision domain case ($r_d^i(0) = 2$) as against a mere 8% in the constant
decision domain case ($r_d^i(t) = 2, \; \forall \, t$). These experiments suggest that the use of a
variable decision domain, instead of a constant one, significantly improves the ability
of the algorithm to capture multiple peaks.

4.3.2   Tests on the Circles Function

The Circles function has connected maxima in the form of concentric circles
(Fig. 4.2). The initial placement consists of random deployment of $n = 1000$ glowworms in the search space $(-10, 10) \times (-10, 10)$. The trajectories resemble the
concentric-circle contours of local maxima, as shown in Fig. 4.18. However, rather
than spreading uniformly on the circles, the glowworms converge to distinct points
on the circles. This mainly occurs as a result of the glowworms seeking to move
toward relatively brighter neighbors even as the associated differences in luciferin
levels may be very small. Some amount of spreading can be achieved by disabling the
attraction between two glowworms whenever the difference in their luciferin values
is less than a small threshold value.

Fig. 4.18 Emergence of the
solution over 250 iterations
for the Circles function

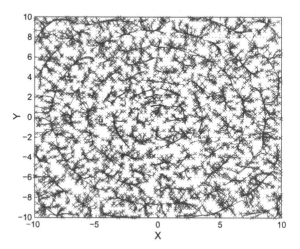

4.3.3 Tests on the Staircase Function

The Staircase function (Fig. 4.3c) has connected local maxima at different levels with discontinuity in the function. It contains 16 regions at 7 different heights in the square $(-2, 2) \times (-2, 2)$. The initial locations of the glowworms, the trajectories followed by the glowworms, and the final solution are shown in Fig. 4.19a, b and c, respectively. Different square regions are marked according to the increasing order of their function values. Note that regions marked with the same number have equal function values. All the glowworms initially located in region 7 remain stationary as they are located in the flat region of global maxima. Note that glowworms in the interior locations (except those that are very close to or on the edges) of all other regions move toward and settle on the edges of the next higher peak-regions. Refer to Fig. 4.19c to observe this pattern.

Even though the Staircase function has step discontinuities, it provides an uphill direction toward the highest level in discrete steps whose size is equal to the width of each stair. Therefore, if the values of the sensor range r_s and of the number of glowworms n are such that there is sufficient connectivity among the swarm members, all the glowworms will reach the highest level. However, the values of r_s and n can be tuned (as is the case in the above simulation result) so that the glowworms detect the presence of other stairs. From a global optimization point of view (i.e., when the goal is to compute only the global peak), it appears that GSO does not perform well for optimization with J_5. However, in the context of this book, since multiple local peaks are being sought, GSO's performance on the J_5 function is indeed good.

4.3.4 Tests on the Plateaus Function

The Plateaus function is used to test the behavior of the algorithm on a profile that has multiple plateaus with equal function values. The plateaus function has connected local maxima at the same level with discontinuity in the function Fig. 4.4d). One thousand glowworms are deployed in the square $(-2\pi, 2\pi) \times (-2\pi, 2\pi)$ (Fig. 4.20a). Figure 4.20b and c show the trajectories followed by the glowworms and the final location of the glowworms, respectively. Note that all the glowworms, with the exception of a very few that are located in the minima regions of the function, settle on the plateau-regions, after the solution is reached. The test results on the J_6 function provide an interesting intuition regarding the working of the algorithm. For instance, we notice from Fig. 4.20b that glowworms located in the interior region of a plateau are always stationary. This could be attributed to their equal luciferin-levels. Note from Fig. 4.20b that most glowworms that climb the gradient steps settle on the plateau edges. However, a few glowworms, after climbing the plateaus, continue moving into the interior regions of the plateaus until their luciferin values reach a steady state and become equal to that of the glowworms located in the interior region.

Fig. 4.19 Test results on the
Staircase function. **a** Initial
placement of glowworms. **b**
Emergence of the solution
after 250 iterations. **c** Final
co-location of groups of
glowworms

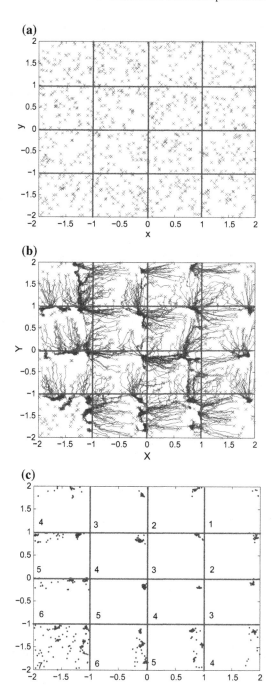

Fig. 4.20 Test results on the
Plateaus function. **a** Initial
placement of glowworms. **b**
Emergence of the solution
over 250 iterations. **c** Final
co-location of groups of
glowworms

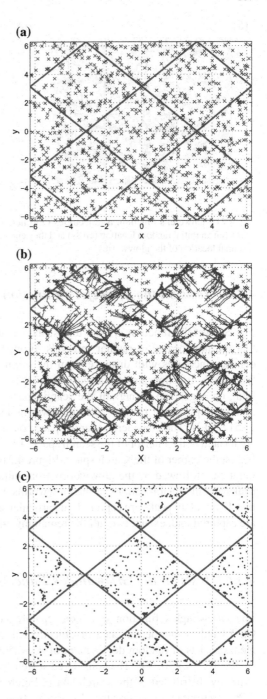

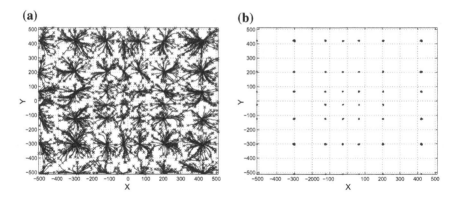

Fig. 4.21 Schwefel's function: **a** trajectories (250 iterations) followed by the glowworms: they start from an initial random location (cross) and they end up at one of the peaks. **b** Initial placement and final location of the glowworms

4.3.5 Tests on the Schwefel's Function

The Schwefel's function is used to test the capability of GSO in capturing peaks that are not located at regular intervals. From Fig. 4.9b, note that the inter-peak distance increases with distance from the origin along each coordinate axis. The Schwefel's function also tests the capability of the algorithm in capturing peaks lying on edges and corners of the search space. The initial placement consists of random deployment of $n = 2000$ glowworms in the search space $(-512, 512) \times (-512, 512)$. The considered search space consists of 49 peaks of which 36 peaks are located in the interior of the search space, 12 peaks lie on the edges, and one peak lies on the corner of the search space. Figure 4.21a and b show the trajectories (250 iterations) followed by the glowworms from initial random locations to one of the peaks and the final locations of the glowworms, respectively. From Fig. 4.21b, note that 47 out of 49 peaks are captured. In particular, all edge-peaks and the corner-peak are captured and, except two, all the remaining 34 interior peaks are captured.

4.3.6 Tests on the Griewangk's Function

The Griewangk's function is a non-separable function and, thereby, tests how GSO handles interdependence of objective function variables. The initial placement consists of random deployment of $n = 2500$ glowworms in the search space $(-30, 30) \times (-30, 30)$. The considered search space consists of 124 peaks that are regularly distributed in the search space. Figure 4.22a and b show the trajectories (500 iterations) followed by the glowworms from initial random locations to one of

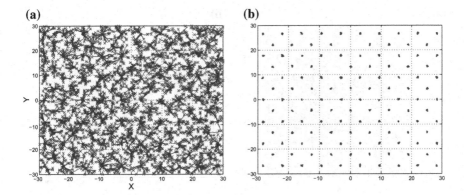

Fig. 4.22 Griewangk's function: **a** trajectories (250 iterations) followed by the glowworms: they start from an initial random location (cross) and they end up at one of the peaks. **b** Initial placement and final location of the glowworms

the peaks and the final locations of the glowworms, respectively. From Fig. 4.22b, note that 122 out of 124 (98.4%) peaks are captured.

4.3.7   Tests on the Ackley's Function

A large-scale view of the Ackley's function in two dimensions looks like a large funnel. The surface of the funnel has a very rough texture with numerous local optima and the global minimum at $(0, 0)$. This property causes the glowworms to move away from the origin, thereby, making it difficult for GSO to find peaks closer to the global minimum. Similar to the Griewangk's function, the Ackley's function is non-separable. Therefore, it also tests how GSO handles interdependence of variables. The initial placement consists of random deployment of $n = 5000$ glowworms in the search space $(-10, 10) \times (-10, 10)$. The considered search space consists of 400 peaks that are regularly distributed in the search space. Figure 4.23a and b show the trajectories (250 iterations) followed by the glowworms from initial random locations to one of the peaks and the final locations of the glowworms, respectively. From Fig. 4.23b, note that 278 out of 400 (70%) peaks are captured. As mentioned above, peaks that are closer to the global minimum are difficult to capture. However, note that the closest peak that is captured is located within a circle of radius 2 units centered around the global minimum.

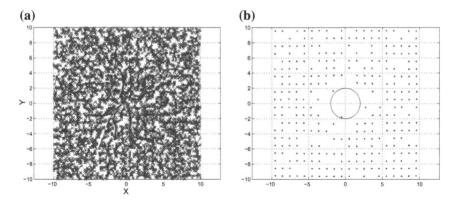

Fig. 4.23 Ackley's function: **a** trajectories (250 iterations) followed by the glowworms: they start from an initial random location (cross) and they end up at one of the peaks. **b** Initial placement and final location of the glowworms

4.3.8 Tests on the Langerman's Function

The Langerman's function is non-symmetrical and its local optima are randomly distributed in the search space. The instance of the Langerman's function shown in Fig. 4.12 is used for this set of experiments. The initial placement consists of random deployment of $n = 5000$ glowworms in the search space $(0, 10) \times (0, 10)$. The function landscape is composed of several circular ridges near the corners $(0, 0)$ and $(10, 10)$ and approximately flat regions on either side of the diagonal formed by the corners $(0, 10)$ and $(10, 0)$. Figure 4.24a and b show the trajectories (250 iterations) followed by the glowworms from initial random locations to one of the peaks and the final locations of the glowworms, respectively. From Fig. 4.24b, note that the all the glowworms move away from the flat regions and settle on the ridges of the search space.

4.3.9 Tests on the Michalewicz's Function

The Michalewicz's function tests how GSO performs on steep valleys and large flat regions with relatively very small slopes (Fig. 4.13). The initial placement consists of a random deployment of $n = 5000$ glowworms in the search space $(0, \pi) \times (0, \pi)$. The considered search space consists of two peaks on its edges and four peaks on its corners. Figure 4.25a and b show the trajectories (500 iterations) followed by the glowworms from initial random locations to one of the peaks and the final locations of the glowworms, respectively. From Fig. 4.25b, note that all the peaks are captured.

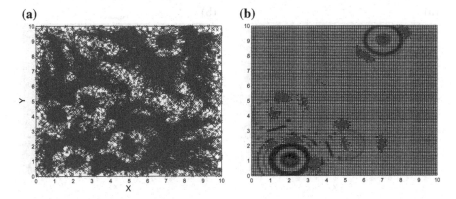

Fig. 4.24 Langerman's function: **a** trajectories (250 iterations) followed by the glowworms: they start from an initial random location (cross) and they end up at one of the peaks. **b** Initial placement and final location of the glowworms

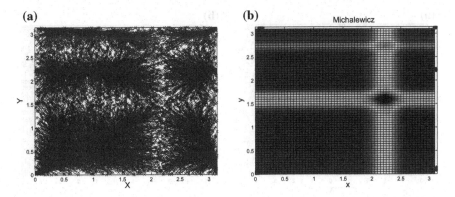

Fig. 4.25 Michalewicz's function: **a** trajectories (250 iterations) followed by the glowworms: they start from an initial random location (cross) and they end up at one of the peaks. **b** Initial placement and final location of the glowworms

4.3.10 Tests on the Drop Wave Function

The Drop wave function tests how GSO performs when the peaks are radially distributed. The initial placement consists of a random deployment of $n = 5000$ glowworms in the search space $(-5.12, 5.12) \times (-5, 12, 5.12)$. The considered search space consists of numerous peaks located radially on circular ridges centered about $(0, 0)$. Figure 4.26a and b show the trajectories (250 iterations) followed by the glowworms from initial random locations to one of the peaks and the final locations of the glowworms, respectively. From Fig. 4.26b, note that as many as 130 peaks are captured.

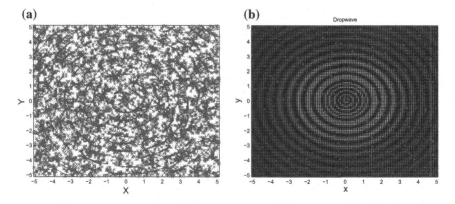

Fig. 4.26 Dropwave function: **a** trajectories (250 iterations) followed by the glowworms: they start from an initial random location (cross) and they end up at one of the peaks. **b** Initial placement and final location of the glowworms

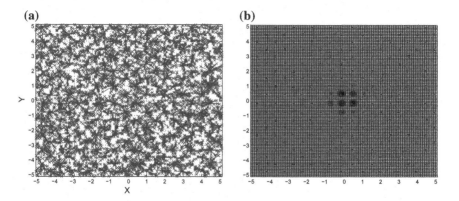

Fig. 4.27 Shubert's function: **a** trajectories (250 iterations) followed by the glowworms: they start from an initial random location (cross) and they end up at one of the peaks. **b** Initial placement and final location of the glowworms

4.3.11 Tests on the Shubert's Function

The Shubert's function tests how GSO performs when the search space contains two sets of peaks with the peak function values in one set significantly different from those of the other. The initial placement consists of a random deployment of $n = 5000$ glowworms in the search space $(-5.12, 5.12) \times (-5, 12, 5.12)$. The considered search space consists of 400 peaks that are regularly distributed in the search space. Figure 4.27a and b show the trajectories (250 iterations) followed by the glowworms from initial random locations to one of the peaks and the final locations of the glowworms, respectively. From Fig. 4.27b, note that 386 out of 400 (96.5%) peaks are captured.

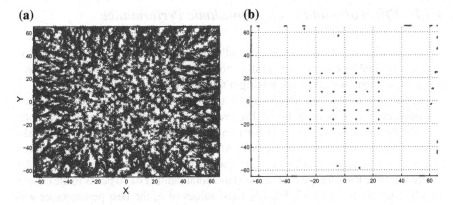

Fig. 4.28 Fifth function of De Jong: **a** trajectories (250 iterations) followed by the glowworms: they start from an initial random location (cross) and they end up at one of the peaks. **b** Initial placement and final location of the glowworms

4.3.12 Tests on the Fifth Function of de Jong

The peaks of the Fifth function of De Jong are arranged in the form of a 2D grid. The initial placement consists of a random deployment of $n = 5000$ glowworms in the search space $(-65.536, 65.536) \times (-65.536, 65.536)$. Figure 4.28a and b show the trajectories (250 iterations) followed by the glowworms from initial random locations to one of the peaks and the final locations of the glowworms, respectively. From Fig. 4.28b, note that the peaks captured by the glowworms are aligned on a grid. Thereby, the glowworms uncover the structure of the Fifth function of De Jong to a large extent.

4.4 Parameter Selection and Performance Evaluation

The number of algorithmic parameters and how much the optimal values of these parameters depend on the problem have a major impact on the practical applicability of optimization algorithms. A good algorithm would consist of a small number of problem-specific parameters and a set of algorithm parameters tuned to fixed values that yield optimal performance over a wide range of problems.

In the set of experiments carried out in Sect. 4.3, fixed values of $\rho, \gamma, \beta, n_t, s,$ and ℓ_0 worked well for all the five function-profile cases. Therefore, as mentioned earlier, these quantities are algorithm parameters that will remain the same and do not need to be specifically tuned for every problem. Thus, only n and r_s need to be selected. Fortunately, their selection is rather straightforward. We show that the choice of these parameters has some influence on the performance of the algorithm, in terms of the total number of peaks captured.

4.4.1 Effect of n and r_s on Algorithmic Performance

In the following set of experiments, the influence of n and r_s on algorithmic performance is studied by carrying out a full factorial analysis on the following functions: Peaks (J_1), Random-peaks (J_2), Rastrigin's (J_3), and Equal-peaks-A (J_7).

In particular, for each combination of n and r_s, the total number of peaks captured is monitored over a set of 30 trials. The average number of peaks captured and the standard deviation values for each function are shown in Tables 4.2–4.5. The Peaks function has a total number of 3 peaks. At $n = 10$, the number of peaks captured increases as r_s increases. When $n \geq 20$, good performance is observed for $r_s \geq 2.5$ (Table 4.2). The Random-peaks function has a total number of 10 peaks in the chosen search space ($[-5, 5] \times [-5, 5]$). For fixed values of n, the best performance was observed for $r_s = 2$ (Table 4.3). The Rastrigin's function has a total number of 16 peaks in the chosen search space ($[-2, 2] \times [-2, 2]$). For fixed values of n, the best performance was observed for values of r_s between 0.5 and 0.75 (Table 4.4). The Equal-peaks-A function has a total number of 9 peaks in the chosen search space ($[-4, 4] \times [-4, 4]$). For fixed values of n, the best performance was observed for

Table 4.2 Results on the Peaks function: Average number of peaks captured, with standard deviation values, for various combinations of n and r_s (30 trials)

$n \vert r_s$	0.5	1.0	1.5	2.0	2.5	3.0	3.5	4.0	4.5	5.0
10	0.0±0.2	0.1±0.3	0.3±0.5	0.4±0.5	0.5±0.5	0.5±0.5	0.6±0.5	0.6±0.5	0.6±0.5	0.7±0.5
20	0.1±0.3	0.4±0.5	0.7±0.7	0.7±0.5	0.9±0.5	0.8±0.5	0.8±0.5	0.8±0.5	0.8±0.5	0.8±0.5
30	0.4±0.6	0.8±0.8	1.3±0.8	1.3±0.6	1.3±0.5	1.2±0.6	1.1±0.5	1.1±0.4	1.1±0.5	1.2±0.5
40	0.6±0.6	1.0±0.7	1.3±0.7	1.5±0.6	1.4±0.5	1.6±0.6	1.4±0.7	1.3±0.5	1.5±0.6	1.4±0.5
50	0.8±0.7	1.4±0.8	1.7±0.7	2.0±0.6	2.1±0.6	2.2±0.7	2.1±0.8	2.2±0.6	2.1±0.6	2.1±0.7
60	0.6±0.7	1.6±0.8	1.8±0.7	2.0±0.5	2.3±0.6	2.1±0.8	2.4±0.6	2.2±0.7	2.3±0.6	2.3±0.6
70	1.0±0.8	1.8±0.7	2.0±0.7	2.2±0.7	2.6±0.6	2.4±0.5	2.4±0.6	2.5±0.5	2.4±0.7	2.5±0.5
80	1.2±0.8	2.0±0.7	2.2±0.5	2.3±0.5	2.6±0.5	2.6±0.5	2.7±0.4	2.5±0.6	2.6±0.6	2.7±0.5
90	1.2±0.8	1.8±0.9	2.3±0.6	2.6±0.6	2.8±0.5	2.7±0.6	2.7±0.4	2.7±0.5	2.8±0.4	2.6±0.5
100	1.5±0.9	2.3±0.6	2.4±0.6	2.5±0.6	2.8±0.4	2.8±0.4	2.8±0.5	2.7±0.5	2.8±0.5	2.8±0.4

Table 4.3 Results on the Random-peaks function: Average number of peaks captured for various combinations of n and r_s (30 trials)

$n \vert r_s$	0.5	1.0	1.5	2.0	2.5	3.0	3.5	4.0	4.5	5.0
50	0.1±0.2	0.4±0.6	1.3±1.0	2.2±1.1	2.3±1.1	2.4±1.0	2.2±1.0	1.9±1.1	1.9±0.6	2.0±0.7
100	0.0±0.2	1.6±1.1	3.6±1.2	3.8±0.9	3.6±0.8	3.5±1.0	3.2±1.0	3.5±1.2	3.4±1.0	3.3±0.9
150	0.2±0.5	3.1±1.3	5.1±1.0	5.4±1.2	4.8±1.0	4.8±0.9	4.6±1.0	4.6±0.8	4.5±1.0	4.6±0.9
200	0.8±0.8	4.6±1.1	6.1±1.1	6.3±0.8	6.0±0.8	5.7±0.9	5.8±0.8	5.4±1.0	5.6±0.8	5.4±1.0
250	1.4±1.0	6.3±1.5	7.1±1.3	7.2±1.3	7.1±1.1	6.6±0.9	6.7±1.1	6.6±1.0	6.4±1.1	6.6±1.3
300	1.7±1.2	6.9±1.3	7.6±0.8	7.5±1.0	7.2±1.0	7.0±1.0	6.9±0.8	6.7±1.2	6.8±1.1	6.6±1.1
350	2.1±1.1	7.6±1.1	8.3±1.0	8.4±1.0	7.9±1.0	7.8±1.1	7.8±1.0	7.6±1.1	7.6±1.0	7.6±0.9

Table 4.4 Results on the Rastrigin's function: Average number of peaks captured, with standard deviation values, for various combinations of n and r_s (30 trials)

$n|r_s$	0.25	0.5	0.75	1.0	1.25	1.5	1.75	2.0
50	0.3±0.5	3.0±1.4	4.6±1.2	4.2±1.1	4.0±0.7	3.5±0.7·	3.4±0.7	3.0±0.8
100	1.4±1.0	7.2±1.5	8.7±1.4	6.6±1.2	5.2±1.1	4.8±0.7	4.7±0.9	4.2±1.2
150	3.8±1.4	11.2±1.9	11.2±1.5	7.6±1.3	6.3±1.1	6.0±0.9	6.1±1.1	6.2±1.1
200	5.6±1.9	13.3±1.3	13.0±1.3	8.2±1.2	8.5±1.9	8.8±1.6	9.4±1.7	8.6±1.7
250	6.8±1.8	14.8±1.1	14.1±1.2	10.1±1.1	12.7±1.3	12.5±1.4	12.2±1.4	11.6±1.7
300	8.4±2.1	15.3±0.8	14.7±1.1	11.6±1.3	14.1±1.1	14.1±1.4	14.4±1.0	14.4±1.2
350	10.4±1.7	15.6±0.6	15.2±0.8	13.2±1.2	15.2±1.1	15.4±0.7	15.6±0.6	15.4±0.7

Table 4.5 Results on the Equal-peaks-A function: Average number of peaks captured, with standard deviation values, for various combinations of n and r_s (30 trials)

$n|r_s$	0.5	1.0	1.5	2.0	2.5	3.0	3.5	4.0
50	0.7±0.9	1.6±1.3	2.6±1.7	2.2±1.7	2.9±1.2	2.4±1.2	1.2±0.9	2.0±0.9
100	2.2±2.8	4.6±2.4	4.3±2.6	5.4±1.8	5.8±1.4	4.8±1.1	3.2±0.9	3.0±1.0
150	3.5±3.8	4.8±2.3	6.8±1.8	7.7±1.4	7.4±1.0	6.0±0.7	4.7±1.2	4.9±1.2
200	2.6±1.1	6.9±1.2	8.1±0.7	8.3±0.6	8.1±0.8	7.3±1.0	6.4±1.0	6.4±1.5
250	3.8±2.3	8.2±0.7	8.6±0.6	8.7±0.5	8.6±0.5	8.1±0.7	7.8±1.0	7.7±0.9
300	5.2±2.4	8.6±0.5	8.8±0.4	8.9±0.3	8.9±0.3	8.7±0.6	8.4±0.6	8.4±0.7
350	4.3±1.3	8.9±0.3	8.9±0.2	8.9±0.2	8.9±0.3	8.6±0.5	8.6±0.5	8.7±0.5

values of r_s between 2 and 2.5 (Table 4.5). For all functions, at any fixed value of r_s, it is observed that the peak capture level increases as n increases.

Four major conclusions can be deduced from the above observations and the results shown in Tables 4.2–4.5.

1. When a small value of r_s relative to the size of the search space is used, a significant fraction of glowworms may get isolated and become stationary at every iteration. Therefore, only a large number of glowworms can ensure connectivity among the swarm members, leading to the computation of all the multiple peaks.
2. For large values of r_s, the number of peaks that influence the movements of a glowworm increases when r_s increases. This may result in a few peak locations being over-populated by glowworms, while other peaks remain undetected at the final iteration, unless a large number of glowworms is used to ensure sufficient allocation of glowworms to all the peak locations.
3. For a fixed value of n, the number of peaks captured increases with the increase in r_s, reaches a maximum at some r_s, and decreases with further increase in r_s.
4. There exists a sensor range r_s^* that minimizes the number of necessary agents without affecting final performance in terms of reached peaks and that is invariant across different peak-capture levels. However, in order to increase the peak-capture level it is necessary to increase the number of glowworms.

From conclusion 4, note that the value of r_s^* is approximately the same for different values of n. Therefore, the value of r_s^* is initially found by keeping n at a lower value that takes relatively less computational time. Then, we set $r_s = r_s^*$ and keep increasing the value of n in order to obtain higher peak-capture levels.

4.4.2   Fixed Algorithm Parameters

The quantities n_t, s, ℓ_0, β, ρ, and γ are algorithm parameters for which appropriate values have been determined based on extensive numerical experiments and are kept fixed (Table 2.1). The neighborhood threshold n_t indirectly controls the number of neighbors of each glowworm by influencing the neighborhood range at each iteration. Whereas a very low value of n_t would not allow enough connectivity for interactions between glowworms, a high value of n_t would result in their strong grouping leading to reduced diversity in the swarm. It was observed that a value of $n_t = 5$ was sufficient to ensure that glowworms are not isolated, yet diversity is maintained between subswarms. The value of step-size s influences the number of iterations in which the peaks are reached by the glowworms and the precision of the solutions. The value of s was selected such that it is very less in relation to the size of the search space. Fixed value of $s = 0.03$ resulted in similar algorithmic performance across all the test functions used in this book. Even though all the glowworms start with the same luciferin value ℓ_0, their luciferin values get updated based on the objective fitness values at their initial positions before they start moving. Therefore, the value of ℓ_0 can be arbitrarily selected. A value of $\ell_0 = 5$ was found to be a good choice. The parameter β affects the rate of change of the neighborhood range. A relatively high value of β would lead to saturation resulting in the switching of the neighborhood range between its upper and lower limits. Therefore, a small value of β (= 0.08) was chosen, which worked well for all the test functions used in this book. A value $\rho = 0$ renders the algorithm memoryless where the luciferin value of each glowworm depends only on the fitness value of its current position. However, $\rho \in (0, 1]$ leads to the reflection of the cumulative goodness of the path followed by the glowworms in their current luciferin values. A value of $\rho = 0.4$ showed good performance across different test functions. The parameter γ only scales the function fitness values and the chosen value of γ (= 0.6) showed good performance for all test functions used in this book.

In the above analysis, the influence of the parameters n and r_s on algorithmic performance has been studied, and synthetic evidence about the fact that algorithmic parameters are indeed constant values has been provided. In conclusion, these results show that n and r_s are the only parameters that need to be selected.

4.5 Tests in Higher Dimensional Spaces

This section deals with the problem of searching higher dimensional spaces using GSO. With an ability to search for local peaks of a function (which is the measure of fitness) in high dimensions, GSO can be applied to identification of multiple data clusters, satisfying some measure of fitness defined on the data, in high dimensional databases. Tests are performed on the Rastrigin's, the Schwefel's, and the Equal-peaks-B functions.

As mentioned earlier, a glowworm is considered to be located at a peak when its position is within a small ϵ-distance from the peak and it is assumed that a peak is captured when at least three glowworms are located at the peak. The value $\epsilon = 0.03$ is chosen for tests conducted on the Rastrigin's function and the Equal-peaks-B function. The value $\epsilon = 0.3$ was used for tests conducted on the Schwefel's function as the considered search space was very large when compared to those of the two former functions. As mentioned earlier, the Schwefel's function makes it difficult for the algorithm to locate peaks on edges and corners of the search space. This is evident in the high dimensional test results that are shown later (See Fig. 4.30).

Results reported from tests conducted up to a maximum of eight dimensions show the efficacy of GSO in capturing multiple peaks in high dimensions. The *average peak-capture fraction* is used as an index to analyze GSO's performance as a function of dimension number. The search space size used in the experiments, the number of peaks in m dimensions for the considered search space size, and the peak locations for $m = 2$ for each function are shown in Table 4.6. From the table, note that each function has 4, 8, 16, 32, 64, 128, and 256 peaks in 2, 3, 4, 5, 6, 7, and 8 dimensions, respectively. The fraction of peaks captured averaged over a set of thirty trials is used as a measure to access the performance of GSO in capturing multiple peaks at each dimension. A maximum value of $n = 10,000$ is used in the experiments. For each multimodal function, the tests are conducted up to a value of m at which the average peak capture fraction drops below 0.5.

Figure 4.29 shows a plot of average fraction of peaks captured as a function of dimension number m when Rastrigin's function is used in the experiments. Note that 100% peak capture is obtained up to $m = 6$, followed by peak captures of $98 \pm 1.3\%$ and $45 \pm 33.8\%$ at $m = 7$ and 8, respectively. Figure 4.30 shows the test results in the case of Schwefel's function. Note that 100% peak capture is found only in two

Table 4.6 Search space size, number of peaks, and peak locations for $m = 2$ for each multimodal function

Function	Search space size	No. of peaks	Peak locations for $m = 2$
Rastrigin's	$\prod_{i=1}^{m} [-1, 1]$	2^m	$(\pm 0.5268, \pm 0.5268)$
Schwefel's	$\prod_{i=1}^{m} [-10, 10]$	2^m	$(-5.24, -5.24)(-5.24, 10),$ $(10, -5.24), (10, 10)$
Equal-peaks-B	$\prod_{i=1}^{m} [-\pi, \pi]$	2^m	$(\pm \frac{\pi}{2}, \pm \frac{\pi}{2})$

Fig. 4.29 Test results on Rastrigin's function: Plot of average fraction of peaks captured as a function of dimension number m (averaged over 30 trials)

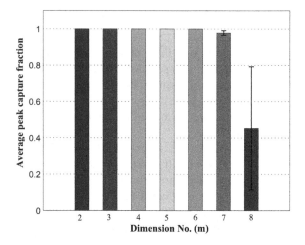

Fig. 4.30 Test results on Schwefel's function: Plot of average fraction of peaks captured as a function of dimension number m (averaged over 30 trials)

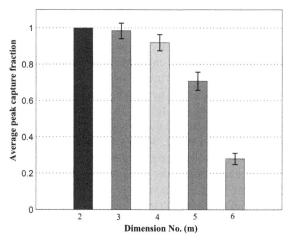

dimensions, followed by 98.3 ± 4.3, $92 \pm 4.4\%$, and $70 \pm 5\%$ at $m = 3$, 4, and 5, respectively. The peak capturing performance degrades from $m = 6$ onwards. As mentioned earlier, in the case of the Schwefel's function, for any dimension m, only one out of 2^m peaks is interior to the search space and each of the remaining peaks is located either on an edge or a corner of the search space making them difficult to be found by the algorithm. This observation can be supported by the results shown in Figs. 4.29 and 4.30. Note that GSO performs well up to seven dimensions in the case of Rastrigin's function. However, good performance is found only up to $m = 5$ in the case of Schwefel's function. The performance of GSO on the Equal-peaks function (Fig. 4.31) is similar to that of the Rastrigin's function up to six dimensions. Note the similarity in the shape of the function profiles of Rastrigin's and Equal-peaks-B functions. However, the considered search space for the latter is larger compared to

Fig. 4.31 Test results on Equal-peaks function: Plot of average fraction of peaks captured as a function of dimension number m (averaged over 30 trials)

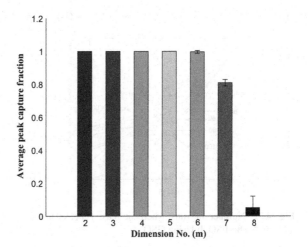

that of the former function. This explains why GSO performs differently on these functions in seven and eight dimensions.

4.6 Experimental Comparison Between PSO and GSO

GSO is compared with Niche-PSO brits:2002, a PSO variant developed for computing multiple optima of multimodal functions. Both the algorithms are tested on Himmelblau's, the Equal-peaks-C, and Rastrigin's functions. Each test is repeated over thirty trials. For the purpose of comparison, the following five performance metrics are used:

1. the number of runs R_i for which at least i ($i = 1, 2, \ldots, Q$) peaks are captured, for various values of n
2. the average number of peaks captured as a function of n
3. the mean distance traveled by an agent
4. the computation time
5. the total number of iterations for convergence.

Tests on Himmelblau's function: Himmelblau's function has four equal peaks of equal function value in the search space of $[-5, 5] \times [-5, 5]$. The plots of R_i for various values of n when Niche-PSO and GSO are used are shown in Fig. 4.32a and b, respectively. For Niche-PSO, the performance improves as the number of agents increases until $n = 40$ (when the number of runs for which all four peaks are captured is maximum ($R_4 = 13$)), after which its performance deteriorates. However, in the case of GSO there is a steady improvement in performance with increasing values of n. The above observation is also supported by the gradual increase in the average number of peaks captured with the increase of n (Fig. 4.33) when GSO is used. The

(a)   **(b)**

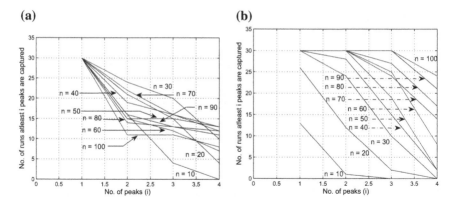

Fig. 4.32 Number of runs where at least i peaks are captured for Himmelblau's function:
a Using Niche-PSO. **b** Using GSO

Fig. 4.33 Average number
of peaks captured as a
function of the number of
agents for Himmelblau's
function

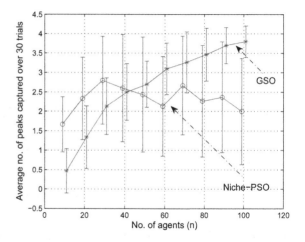

maximum average number of peaks captured is higher when GSO is used than when
Niche-PSO is used. Moreover, the standard deviation values are relatively less in the
case of GSO than those of PSO.

Tests on the Equal-peaks-C function: The Equal-peaks-C function has 12 peaks of
equal function value in the search space of $[-5, 5] \times [-5, 5]$. The plots of R_i for
various values of n when the Niche-PSO and GSO are used are shown in Fig. 4.34a
and b, respectively. Similar comparison results are obtained as in the experiments with
Himmelblau's function. In the case of Niche-PSO, note that performance degrades
drastically above $n = 100$ (Fig. 4.34a), where only one peak is captured (with the
exception of two peak captures in a few instances). The maximum average number
of peaks captured is only 2 in the case of Niche-PSO and is approximately 12 in the
case of GSO (Fig. 4.35).

(a)

(b)

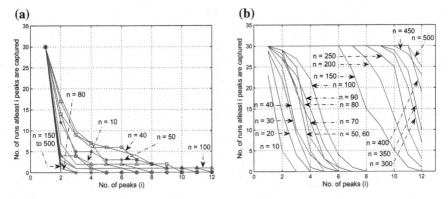

Fig. 4.34 Number of runs where at least *i* peaks are captured for the Equal-peaks-B function: **a** Using Niche-PSO. **b** Using GSO

Fig. 4.35 Average number of peaks captured as a function of the number of agents for the Equal-peaks-B function

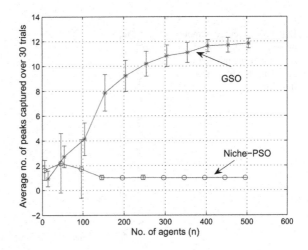

Tests on Rastrgin's function: Rastrigin's function consists of 16 peaks of unequal function values in a search space of $[-2, 2] \times [-2, 2]$. The plots of R_i for various values of n when the Niche-PSO and GSO are used are shown in Fig. 4.36a and b, respectively. The average number of peaks captured as a function of n is shown in Fig. 4.37. Using the same analysis as before, it is evident that GSO performs better when compared to Niche-PSO on Rastrigin's function.

A common observation in all the above experiments is the fact that Niche-PSO maintains disjoint sub-swarms at low values of n during its search for multiple peaks, but the performance in terms of number of peaks captured is not consistent over the number of trials. Moreover, as the number of agents increases, the algorithm loses the niching ability, leading to the convergence of all agents to a single peak at relatively large values of n. However, GSO shows stable performance over different trials and shows a steady improvement in performance with increasing values of n in terms

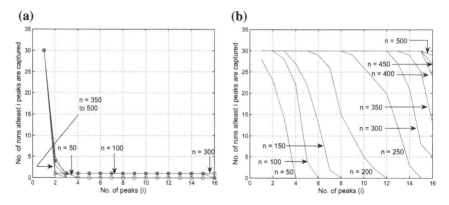

Fig. 4.36 Number of runs where at least *i* peaks are captured for Rastrigin's function: **a** Using Niche-PSO. **b** Using GSO

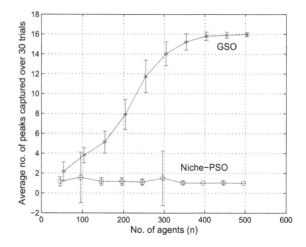

Fig. 4.37 Average number of peaks captured as a function of the number of agents for Rastrigin's function

of the average number of peaks captured. The mean distances traveled by an agent for different cases are plotted in Fig. 4.38a, b and c. Note that the average distance traveled by an agent is relatively very low for GSO in comparison to Niche-PSO, in all the cases. Even though GSO takes more time than Niche-PSO at higher values of *n* (Fig. 4.39), GSO performs better than Niche-PSO in terms of the number of peaks captured.

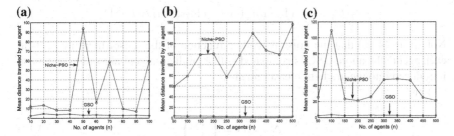

Fig. 4.38 Mean distance travelled by an agent in the case of different test functions: **a** Himmelblau's. **b** Equal-peaks. **c** Rastrigin's

Fig. 4.39 Total computation time for 300 iterations

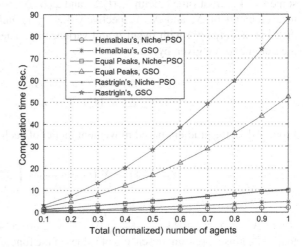

4.7 Performance in the Presence of Noise

Results demonstrating the capability of GSO algorithm to capture multiple peaks of a number of complex multimodal functions are reported in Sect. 4.3. In this section, the algorithm's performance in the presence of noise is reported. The Peaks function J_1, given in (2.24), is considered for this purpose.

4.7.1 Modelling of Measurement Noise

The behavior of the algorithm is examined in the presence of uncertainty due to measurement noise. A Gaussian distribution $N(\mu, \sigma_f^2)$ with mean μ and variance σ_f^2 is used to model the sensor noise $\omega(t)$. Since the noise introduced affects the quality of measurements of the function values, the luciferin-update rule is modified to take into account the effect of noise as follows:

$$\ell_j(t+1) = (1-\rho)\ell_j(t) + \gamma(1+w(t))J_j(t+1) \tag{4.16}$$

Note from (4.16) that the noise is made a function of the sensor measurements at each agent's location. GSO is compared with a gradient-ascent algorithm and noise introduced into the gradient measurements:

Let

$$\Delta_{(x,y)}(t) = \begin{pmatrix} \Delta_x \\ \Delta_y \end{pmatrix} \tag{4.17}$$

represent the gradient vector measured at the location (x, y) by a glowworm i for some $i = 1 \ldots n$ at time t, with $w_x(t)\Delta_x$ and $w_y(t)\Delta_y$ being the noise levels in the measurements of Δ_x and Δ_y, respectively. Let $w_x(t)$ and $w_y(t)$ follow a Gaussian distribution $N(\mu, \sigma_g^2)$ with mean μ and variance σ_g^2. Therefore, the noisy gradient measurements are given by:

$$\begin{aligned} \tilde{\Delta}_x &= (1 + w_x(t))\Delta_x \\ \tilde{\Delta}_y &= (1 + w_y(t))\Delta_y \end{aligned} \tag{4.18}$$

From (4.18), the gradient based movement model of each glowworm i is given by:

$$X_i(t+1) = X_i(t) + \frac{\delta}{|\tilde{\Delta}_{(x,y)}|} \begin{pmatrix} \tilde{\Delta}_x \\ \tilde{\Delta}_y \end{pmatrix} \tag{4.19}$$

where, $X_i = (x_i\ y_i)$ and $|\tilde{\Delta}_{(x,y)}| = \sqrt{\tilde{\Delta}_x^2 + \tilde{\Delta}_y^2}$. From (4.16) and (4.18), it is clear that while noise is present in sensing of values of the function profile in the case of the GSO algorithm, noise occurs in gradient measurements in the latter case. Therefore, the comparison of performance between the two algorithms is not straightforward. For this purpose, a simple relation between the variances involved is derived in both the cases. Figure 4.40a shows the variation of J with respect to x. The quantities $\Delta J_{\min}$ and $\Delta J_{\max}$ represent the minimum and maximum values of $\tilde{\Delta}J$ for a corresponding value of $\Delta x = x_2 - x_1$, where $\tilde{\Delta}J$ represents the difference between noisy function values measured at x_1 and x_2. From the figure we have,

$$\begin{aligned} \Delta J_{\min} &= J_2 - J_1 - \sigma_f(J_1 + J_2) \\ \Delta J_{\max} &= J_2 - J_1 + \sigma_f(J_1 + J_2) \end{aligned} \tag{4.20}$$

$$\Rightarrow \frac{\tilde{\Delta}J}{\Delta x} \in \left(\frac{\Delta J}{\Delta x} - \sigma_f \frac{J_1 + J_2}{\Delta x}, \frac{\Delta J}{\Delta x} + \sigma_f \frac{J_1 + J_2}{\Delta x} \right) \tag{4.21}$$

Assuming $J_1 \approx J_2$ for small variation in the function values with respect to small deviation in x, we get

Fig. 4.40 **a** Graph of J
versus x is used to find the
relationship between
variance in the function
values and the variance in the
gradients computed by using
a finite-difference method. **b**
$\pm\delta_1$ and $\pm\delta_2$ represent the
angular dispersions in the
directions of the nominal
gradient when the same
amount of noise is
introduced into the function
values (GSO algorithm) and
gradient values (gradient
algorithm), respectively

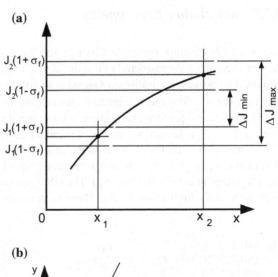

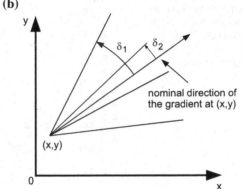

$$\sigma_g \approx \left(\frac{2J_1}{\Delta x}\right)\sigma_f \qquad (4.22)$$

From (4.22), it is clear that a small variance in the function value leads to a very
large variance in gradient value obtained by a finite-difference method, especially as
$\Delta x \to 0$. In the simulations, variance values in the gradient algorithm that are of the
same order as the variance values in the GSO algorithm are used. This means that
far less noise in the gradient algorithm is dealt with than in the GSO algorithm. The
gradient-cone interpretation given in Fig. 4.40b explains the above fact in the following manner. Here, $\pm\delta_1$ and $\pm\delta_2$ represent the angular dispersions in the directions
of the nominal gradient when the same amount of noise is introduced into the function values (GSO algorithm) and gradient values (gradient algorithm), respectively.
Clearly, $\delta_1 > \delta_2$. Note that for small values of Δx, the angle of the outer cone can
actually become much larger, leading to the condition $\delta_1 \gg \delta_2$.

4.7.2 Simulation Experiments

The J_1 (2.24) function (given in Chap. 2) and the noise models, described in the
previous section, are considered to test the algorithmic response to uncertainty conditions. The noise $\omega(t)$ follows a Gaussian distribution $N(0, \sigma_f^2)$ with zero mean
and variance σ_f^2. The mean minimum distance to the sources $d_{min,av}(t)$, given in
(4.15), is used as a performance metric in these simulations. Figure 4.41a shows the
plots of $d_{min,av}(t)$, for several values of standard deviation σ_f, averaged over a set
of 20 simulation trials for each σ_f, in the case of the GSO algorithm. The standard
deviation of $d_{min,av}(t)$, in each case, is relatively very less when compared to that of
the introduced noise ($\approx 0.02 \times \sigma_f$). The GSO algorithm shows good performance
even with fairly high noise levels. There is graceful degradation of performance

Fig. 4.41 Plots of $d_{min,av}(t)$
(averaged over 20 trials) for
a various values of σ_f, in the
case of GSO algorithm. **b**
Various values of σ_g, in the
case of gradient based
algorithm

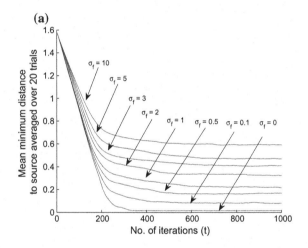

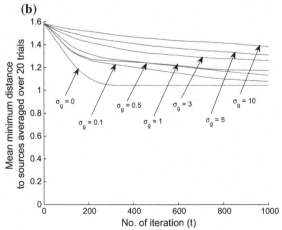

only with significant increase in levels of measurement noise. However, gradient based algorithm degrades rather fast in the presence of noise (Fig. 4.41b). Note that the behavior of $d_{min,av}$ in the case of GSO algorithm, with $\sigma_f = 10$, is relatively better than the behavior of $d_{min,av}$ in the case of gradient based algorithm, with $\sigma_g = 0.1$. Multimodal optimization could also be achieved by distributing the agents uniformly in the solution space and allowing them to perform gradient descent to the local peaks, using no communication at all. However, notice that a random initial placement results in most of agents moving away from the peaks and settling at the edges of the solution space. This is evident in Fig. 4.41b where the mean minimum distance to the peaks $d_{min,av}(t)$ does not reduce to zero even in the absence of noise ($\sigma_g = 0$).

4.8 Summary

In this chapter, the GSO algorithm was evaluated on several standard multimodal test functions borrowed from the literature. By choosing non-differentiable functions such as the stair-case and multiple-plateau functions, it was shown that the algorithm can handle discontinuities in the objective function. Test results in higher dimensions with large number of peaks were reported. Some experimental comparisons of GSO with a PSO variant used for capturing multiple peaks was presented. The approach of GSO to multimodal optimization problems raises several intriguing questions about the relation of parameter values and tuning of the algorithm based on different applications. Some of these questions were answered by providing values for the algorithm parameters that were found based on experimental observations and by showing that the values chosen worked well for a wide range of experimental scenarios. Moreover, experiments were conducted to show that only two parameters, n and r_s, influence algorithmic performance and need to be selected by the user. The experimental study in this regard gives a profound insight into selecting these two parameters for any problem. Future work in this direction would involve a thorough analytical study of the effect of various parameters on algorithm performance, aimed primarily toward providing an analytical justification to the conclusions reached by experimentation. Finally, algorithmic behavior was examined in the presence of noise. Simulations were used to show that the algorithm exhibits good performance in the presence of fairly high noise levels. Graceful degradation was observed only with significant increase in levels of measurement noise. A comparison between the glowworm algorithm and the gradient based algorithm revealed the superiority of GSO in coping with uncertainty. GSO was originally developed for numerical multimodal function optimization problems. However, it could be applied to a class of robotics tasks related to localization of multiple signal sources. In the next chapter, embodied simulations and real-robot-experiments are used to demonstrate the application of GSO to source localization problems.

4.9 Thought and Computer Exercises

Exercise 4.1 Consider the Rastrigin's function (4.1) in two dimensions ($m = 2$) in a search space of $[-3, 3] \times [-3, 3]$:

$$J(x, y) = 20 + x^2 + y^2 - 10\cos(2\pi x) - 10\cos(2\pi y)$$

4.1.1 Show that the optima of this function cannot be found analytically (Hint: Find the expressions for the partial derivatives $\frac{\partial J}{\partial x}$, $\frac{\partial J}{\partial y}$ and show that the first-order conditions take the form $f(\theta) = A\theta + \sin(B\theta) = 0$, which cannot be solved exactly).

4.1.2 To make an initial guess of how many peaks exist and where the peaks may lie, plot the function in the chosen search space and count the number of peaks manually (Use the PlotFunction.m for this purpose. Every red spot can be counted as a peak). Verify by inspection of this plot that the function has peaks in the vicinity of the following locations $Q_i = \left(\pm(\frac{1}{2} + i), \pm(\frac{1}{2} + i)\right), i = 1, 2, \ldots$ Do the peaks lie exactly at these locations? (Hint: Analyze the first order conditions in 4.1.1 for these locations).

4.1.3 In the following experiments, overlay your plots with a small circle at each of the locations Q_i to evaluate the number of peaks captured. Set $n = 500$ and $r_s = 2$. Use Table 2.1 to set the values for fixed algorithmic parameters. Similar to Exercise 2.6, run the GSO code and record the number of peaks captured for the constant neighborhood and adaptive neighborhood cases, respectively. What do you observe?

Exercise 4.2 Until now, you determined if a peak is captured or not by just looking at the plot of final locations of the glowworms and checking if at least three glowworms were located very close to Q_i. Write a routine to perform this step automatically (Hint: Consider each Q_i and compute its distance to final location of each glowworm. If this distance to the glowworm is less than ϵ (= 0.05), then that glowworm is co-located at a corresponding peak. Extend this routine to count the total number of peaks captured. Embed this routine into the GSO code and repeat Exercise 4.1.3. Verify if the routine works properly by comparing the results with those that were obtained manually in the previous exercise.

Exercise 4.3 Monte Carlo methods depend on repeated random trials to extract any statistically significant behavior in the performance of an algorithm. For this purpose, set up the code to run $N(= 30)$ number of random trials of the tests in Exercise 4.1. Record the number of peaks captured in each trial. After all the trials are over, compute mean and standard error of the mean (SEM) of the number of peaks captured in each case. Comment on your observations. Repeat the same experiments for an additional 20 times and obtain the mean and SEM values for $N = 50$. Next, repeat the same experiments for an additional 50 times and obtain the mean and SEM values for $N = 100$. Comment on how the results change with an increase in the sample size.

Exercise 4.4 Consider the Equal-peaks-A function (4.5) in two dimensions:

$$J(x, y) = \cos^2(x) + \cos^2(y)$$

4.4.1 Show analytically that the peaks occur at $Q_{ij} = (\pm i\pi, \pm j\pi), i = 0, 1, \ldots;$ $j = 0, 1, \ldots$ (Hint: Show that the Kuhn–Tucker (KT) first-order necessary conditions and second-order sufficient conditions are satisfied at these locations. Please refer to a book on optimization to learn about KT conditions).

4.4.2 Consider a search space of $[-8, 8] \times [-8, 8]$. In the following experiments, overlay your plots with a small circle at each of the locations Q_{ij} to evaluate the number of peaks captured. Set $n = 200$ and $r_s = 3$. Use Table 2.1 to set the values for fixed algorithmic parameters. Run the GSO code and count the number of peaks captured. Increase n in steps of 50 up to 500 and record the number of peaks captured in each case. Comment on your observations.

4.4.3 Run 30 trials of the experiments in 4.4.2 and record the mean and SEM values in each case. Use these results to plot the graph of average number of peaks captured as a function of n. Also plot the Mean $\pm$ SEM values on this graph. Comment on your observations.

Exercise 4.5 Consider the Equal-peaks-B function (4.6) in two dimensions:

$$J(x, y) = \sin^2(x) + \sin^2(y)$$

4.5.1 Show analytically that the peaks occur at $Q_{ij} = \left(\pm(2i + 1)\frac{\pi}{2}, \pm(2j + 1)\frac{\pi}{2}\right),$ $i = 0, 1, \ldots; j = 0, 1, \ldots$.

4.5.2 Consider a search space of $[-6, 6] \times [-6, 6]$. In the following experiments, overlay your plots with a small circle at each of the locations Q_{ij} to evaluate the number of peaks captured. Set $n = 200$ and $r_s = 3$. Use Table 2.1 to set the values for fixed algorithmic parameters. Run the GSO code and count the number of peaks captured. Increase n in steps of 50 up to 500 and record the number of peaks captured in each case. Comment on your observations.

4.5.3 Run 30 trials of the experiments in 4.5.2 and record the mean and SEM values in each case. Use these results to plot the graph of average number of peaks captured as a function of n. Comment on your observations.

Exercise 4.6 Consider the Equal-peaks-C function (4.7) in two dimensions:

$$J(x, y) = \cos^2(x) + \sin^2(y)$$

4.6.1 Show analytically that the peaks occur at $Q_{ij} = \left(\pm i\pi, \pm(2j + 1)\frac{\pi}{2}\right), i = 0, 1, \ldots; j = 0, 1, \ldots$.

4.6.2 Consider a search space of $[-7, 7] \times [-7, 7]$. In the following experiments, overlay your plots with a small circle at each of the locations Q_{ij} to evaluate the number of peaks captured. Set $n = 200$ and $r_s = 3$. Use Table 2.1 to set the values for fixed algorithmic parameters. Run the GSO code and count

the number of peaks captured. Increase n in steps of 50 up to 500 and record
the number of peaks captured in each case. Comment on your observations.

4.6.3   Run 30 trials of the experiments in 4.6.2 and record the mean and SEM
values in each case. Use these results to plot the graph of average number of
peaks captured as a function of n. Comment on your observations.

Exercise 4.7 Inter-peak distance can be defined as the distance between any two
neighboring peaks. Compute the inter-peak distances for Equal-peaks-A, Equal-
peaks-B, and Equal-peaks-C functions. Are they same through out the search space
for each function? If not, compute the mean and SEM of all the inter-peak distances
in the given search spaces in the previous three exercises. Repeat the same for the
Rastrigin's function in Exercise 4.1. What do you observe? (Does the inter-peak
distance vary a lot in the given search space?).

Exercise 4.8 From Exercises 4.1 to 4.6, certain values of n and r_s were selected to
test GSO on each function. Based on this experience, fix a suitable value of n (neither
too low nor too high) in the case of each function. Now select different values of r_s
appropriately (for example, if you set $r_s = 3$ for testing a particular function, then
select 10 values between 2 and 4); run 30 trials of GSO code for each case and record
the mean and SEM values of number of peaks captured. Obtain a graph of mean
peaks captured vs r_s. Use this graph and results in Exercise 4.7 to comment on the
relationship between the average inter-peak distance and r_s (For example, what ratio
of these two quantities maximizes the number of peaks captured?).

Exercise 4.9 Consider the Schwefel's function (4.8) in two dimensions:

$$J(x, y) = 418.9829 \times 2 + x \sin(\sqrt{|x|}) + y \sin(\sqrt{|y|}) \qquad (4.23)$$

One of the complexities of this function is that the peaks are not located at regular
intervals. Consider a search space of $[-512, 512] \times [-512, 512]$. Set $n = 500, r_s =$
100, and $s = 0.3$. Use Table 4.1 to set the remaining algorithm parameters. Run the
GSO code for 1000 iterations and obtain the plot of final location of glowworms.
Overlay this plot with a contour plot (Use PlotFunction.m for this purpose). Verify
if the glowworms' final locations coincide with the peaks represented by the contour
plot. How many of the peaks are captured? Do you find any glowworms co-located
away from the peaks?

Exercise 4.10 Given the final locations of glowworms and assuming that each glowworm subgroup is co-located at a peak, devise a routine to automatically find the
number and locations of the peaks (Hint: Initialize a new subgroup SG_1 with the
index of first glowworm. Find all the glowworms that fall within a threshold distance
2ϵ of this glowworm and assign them to this subgroup. Assign the centroid of all the
glowworms in this subgroup to a corresponding peak location. Repeat this procedure
until all the glowworms have been assigned to one of the subgroups). Embed this in
the GSO code and repeat the experiment in Exercise 4.9 (Set $\epsilon = 0.5$). Verify that the
results match with those obtained in the previous exercise. Now that the numerical

values of the peaks have been found, evaluate the inter-peak distances and compute the corresponding mean and SEM values.

Exercise 4.11 Solving Exercise 4.10 enables automating the running of GSO code for multiple trials, when the peaks are not known beforehand for reference purposes. Use this method to run 30 trials of the experiment in Exercise 4.10 and record: (1) number of peaks, (2) peak locations, and (3) mean inter-peak distance in each case. Compute the mean and SEM values of the three quantities over these 30 trials. Comment on your observations.

Chapter 5
Experiments Using Physical Simulations and Real Robots

In the previous chapter, test results from numerical simulations were reported to validate GSO's approach to multimodal function optimization. In this chapter, the potential of GSO for signal source localization is demonstrated by using physically realistic simulations and experiments with real robots.

The swarm robotics based approach to source localization usually involves a swarm of mobile robots that search a given area for previously unknown signal-sources. These robots use cues such as their perception of signal-signatures at their current locations, and any information available from their neighbors, in order to guide their movements toward, and eventually converge at, the signal-emitting sources. The GSO algorithm, originally designed for solving numerical optimization problems, serves as an effective swarm robotics approach to the source localization problem.

Certain algorithmic aspects need modifications while implementing in a robotic swarm mainly because of the point-agent model of the basic GSO algorithm and the physical dimensions and dynamics of a real robot. The modifications incorporated into the algorithm in order to make it suitable for a robotic implementation are described. Physical simulations[1] are conducted, by using a multi-robot-simulator called Player/Stage [62] that provides realistic sensor and actuator models, in order to assess GSO's suitability for multiple source localization tasks. Next, this work is extended to robotics experiments. For this purpose, wheeled robots, called Kinbots [111], that were originally built for experiments related to robot formations are used. By making necessary modifications to the Kinbot hardware, the robots are endowed with the capabilities required to implement the various behavioral primitives of GSO. Results from sound source localization [104] and light source localization experiments [108] demonstrate the potential of robots using GSO for localizing signal sources.

---

[1] Simulations whose behavior is constrained by models of physical laws [139].

© Springer International Publishing AG 2017

K.N. Kaipa and D. Ghose, *Glowworm Swarm Optimization,*
Studies in Computational Intelligence 698, DOI 10.1007/978-3-319-51595-3_5

5.1 Physical Simulations

The use of physical simulations offers a quick and efficient way for feasibility testing and has the potential to save time and reduce cost in the development of the real robotic system. They also serve as a tool to perform off-line optimization of task performance across various algorithmic parameters (e.g., group size) before validation with real robots. There are mainly three issues that arise during a robotic implementation that are not taken into account in the algorithmic description of GSO:

1. The linear and angular movements of agents in the algorithm are instantaneous in nature. However, physical robots spend a finite time in performing linear and angular movements based on their prescribed speeds and turn rate capabilities.
2. In the algorithm, agents are considered as points and agent collisions are ignored. These are acceptable assumptions for numerical optimization problems. However, real robots have a physical shape and foot print of finite size, and cannot have intersecting trajectories. Thus, robots must avoid collisions with other robots in addition to avoiding obstacles in the environment.
3. GSO uses a leapfrogging behavior (described in Chap. 2), where agents move over one another, to perform local search. Since robots perform collision avoidance with other robots, a physical implementation of the leapfrogging effect becomes a critical issue and is not straightforward.

The above issues are very important and need a careful consideration as they call for changes in the agent movement models and alter the working of the basic algorithm when it is implemented on a swarm of real robots. For instance, since physical robots cannot move over one another, a direct implementation of the leapfrogging behavior is not possible. However, the obstacle avoidance feature alters the algorithm's behavior, giving rise to a partial leapfrogging effect, which is described in the next subsection. Therefore, the physical simulation experiments that are presented in this chapter serve to achieve several goals such as modeling the algorithmic modifications needed for robotic implementation, examining the altered algorithm's behavior, and studying the robustness of the algorithm to implementational constraints.

5.1.1 GSO Variant for Physical Simulations

Each member of the mobile robot swarm used to implement GSO should possess the following capabilities:

1. Sensing and broadcasting of profile-value (luciferin level) at its location.
2. Detection of the number of its neighbors and their locations relative to its own location.
3. Reception of profile-values (luciferin levels) from its neighbors.
4. Selection of a neighbor using a probability distribution (based on the relative luciferin levels of its neighbors) and making a step-movement toward it.

5. Variation of the neighborhood range.
6. Avoiding collisions with obstacles and other robots in the environment.
7. Leapfrogging behavior for performing local search.

As a real-robot simulation platform is used and perfect sensing and broadcast is assumed in the initial experiments, the first four robot capabilities are rather straightforward to implement. However, the mechanisms of implementing obstacle avoidance and leapfrogging behaviors are non-trivial and are described below.

Obstacle avoidance: The footprint of the robot is considered to be an octagon with a circumcircle radius r_{robot}. A pair of sonar based proximity-detection sensors with a range r_{sonar} and field of view θ_{fov} are mounted in the frontal region of the robot at a distance d_{sonar} from the robot-center. A simple obstacle avoidance rule is used where the robot decides to turn right (left) if the left (right) sensor detects the other robot. The robot performs the collision-avoidance maneuver by moving along an arc, whose radius of curvature is r_{cv}, and never reaching closer than a safe-distance d_{safe} to the other robot/obstacle. This situation is shown in Fig. 5.1. Using simple geometry, the radius of curvature r_{cv} can be calculated as below:

$$r_{cv} = \frac{(d_{sonar} + r_{sonar} + r_{robot})^2 - (2r_{robot} + d_{safe})^2}{2(r_{robot} + d_{safe})} \tag{5.1}$$

Leapfrogging behavior: The leapfrogging effect of GSO requires a glowworm to move over another glowworm whenever two glowworms are within a step distance s of each other. This is illustrated in Fig. 5.2a where glowworm A moves over glowworm B to reach the position $A(t+1)$. However, this is not directly realizable with physical robots. In the following, three methods to achieve an explicit robotic implementation of the leapfrogging behavior are described. The first method (Fig. 5.2b) involves interchange of agent-roles, through communication, where robot B moves to the desired position of robot A and robot A replaces robot B's position. The illus-

Fig. 5.1 The obstacle avoidance model used by the robots in physical simulations of the GSO algorithm

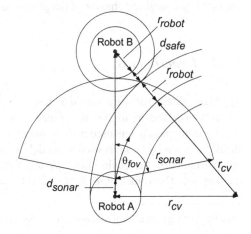

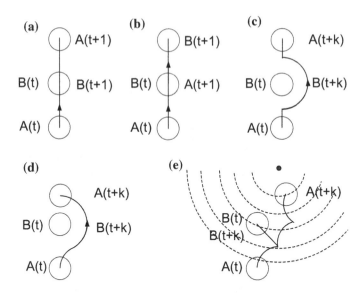

Fig. 5.2 **a** Leapfrogging behavior according to the GSO algorithm. **b**, **c** and **d** Three methods of robotic implementation of explicit leapfrogging behavior. **e** Implicit leapfrogging behavior due to obstacle avoidance

tration in Fig. 5.2c shows another method where the robot A takes a detour about robot B's position and reaches its desired position. As this method is difficult to implement, a modification is proposed in which the detour is achieved using a blend of two circular paths as shown in Fig. 5.2d. It is also observed that the obstacle avoidance behavior automatically gives rise to an implicit leapfrogging effect, which is described using the illustration in Fig. 5.2e. Robot A alternatively performs collision avoidance with, and seeks to move toward, robot B until it crosses the equi-contour line on which robot B is situated. Thereafter, since robot A becomes a leader to robot B, robot B performs similar movements with respect to robot A, thus leading to an implicit leapfrogging behavior of the glowworm-pair.

5.1.2 Experiments

The Player/Stage software is used to carry out the physical simulation experiments.[2] The robot parameters used in the simulations are shown in Table 5.1. The values of the algorithm parameters used in the simulations are shown in Table 5.2. The $J_2(x, y)$ function, given by (3.1), is used to model the multiple signal sources (Fig. 5.3) that are distributed in the environment. A set of three peaks ($Q = 3$) is considered for the purpose. The values of a_i, b_i, x_i, y_i, for $i = 1, \ldots, 3$ are shown in Table 5.3.

---

[2]The Player/Stage simulation results in this section are courtesy Varun R. Kompella.

Table 5.1 Robot parameters

r_{robot}	d_{sonar}	r_{sonar}	r_{cv}	d_{safe}	V_{linear}
8 cm	4 cm	75 cm	15 cm	58.6 cm	0.4 m/s

Table 5.2 Values of algorithm parameters used in the simulations

r_s	ρ	γ	β	n_t
8	0.4	0.6	0.08	2

Fig. 5.3 Model of the multiple signal-source profile used for the physical simulations

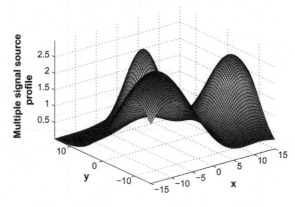

Table 5.3 Values of a_i, b_i, x_i, and y_i used to generate the function profile $J_2(x, y)$

	1	2	3
a_i	3	2.5	2.5
b_i	0.01	0.04	0.02
x_i	−10	0	12
y_i	−10	7	−5

A workspace of size 30×30 sq. m ($[-15, 15] \times [-15, 15]$) is considered and 20 glowworms are deployed in a square region $[-4, 4] \times [-9, -1]$ of size 8×8 sq. m. The center of the workspace is chosen as the center of the circle circumscribing the triangle formed by the three source locations in order to avoid bias of robot movements toward any single source. Figure 5.4a–j show the various stages of an experiment where the swarm of 20 robots split into three subgroups, simultaneously taxis toward, and co-locate at the three source locations. It can be observed that the initial group of robots split into networked subgroups within the first 5 s. These subgroups now taxis toward the source locations in the remaining time. The implicit leapfrogging effect due to obstacle avoidance behavior is clearly apparent in these simulations. Note that the robots do not actually converge to the source locations in an exact sense, but hover around the source due to the collision avoidance feature. This is acceptable in a realistic scenario.

The same experiment is then repeated over a set of 30 trials and performance measures such as the distance traveled by each robot, number of sources captured, and number of robots converging to each source are recorded. The same initial deployment of robots for the first set of 15 trials and different random initial deployments

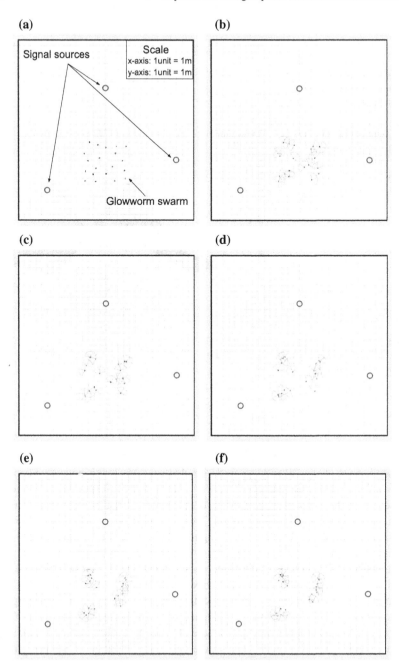

Fig. 5.4 A swarm of 20 glowworms deployed in the region ($[-4, 4] \times [-9, -1]$) split into three subgroups, taxis toward, and localize three sources. The snapshots are shown at **a** $t = 0$ **b** $t = 3$ **c** $t = 3.5$, and **d** $t = 4$ s of simulation time. The snapshots are shown at **a** $t = 4.5$ **b** $t = 5$ **c** $t = 10$, and **d** $t = 15$ s of simulation time. The snapshots are shown at **i** $t = 30$ and **j** $t = 50$ s of simulation time

(g) **(h)**

(i) **(j)**

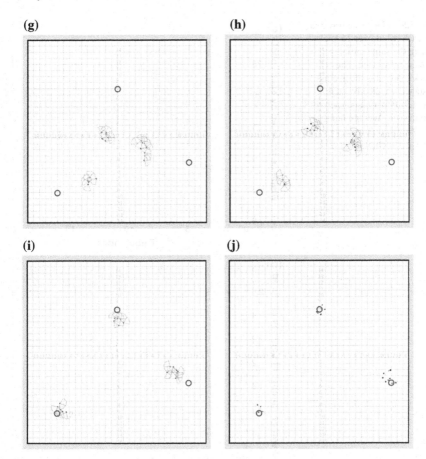

Fig. 5.4 (continued)

for the next set of 15 trials are considered, respectively. Figure 5.5a shows the plot of distance traveled by each robot for the first 15 trials, along with the average behavior. An average of 5, 6, and 9 robots converge to the sources $(-10, -10)$, $(0, 7)$, and $(12, -5)$, respectively, over the 15 trials of same initial placements. In the second set of 15 experimental trials where different initial placements were used, the number of robots converging to each source and the number of sources captured were recorded, and are shown in Table 5.4. In the final set of experiments, the task performance is assessed by monitoring average number of sources captured as a function of the robot group size (Fig. 5.5b).

Fig. 5.5 **a** Distance traveled by each robot over 15 trials where the same initial placement is used. The '∗' mark represents the average distance traveled by each robot. **b** Performance across group size: Average number of sources captures a set of 10 experimental trials

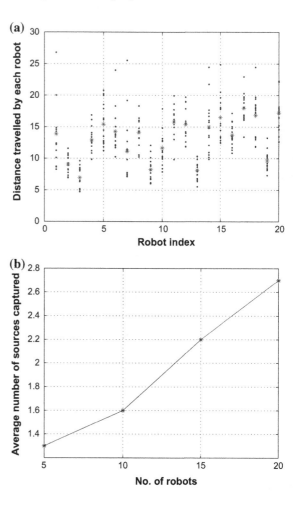

5.2 Glowworms: Multi-robot Platform

Four small-wheeled robots christened Glowworms (named after the glowworm algorithm; see Fig. 5.6) were built upon Kinbots[3] [111] to conduct localization experiments. Realization of each sensing-decision-action cycle of the GSO algorithm requires the robots to perform subtasks such as identification and relative localization of neighbors, selection of a leader among current neighbors, updating of the associated luciferin and local-decision range, and making a step-movement toward the selected leader. By making necessary modifications to the Kinbot hardware, the robots are endowed with the capabilities required to implement the various behavioral primitives of GSO. The first version of Glowworms (Fig. 5.7a) were used for sound

---

[3] Kinbots were originally built for experiments related to robot formations.

Table 5.4 The first three rows show the number of robots converging to each source and the last row shows the number of sources captured over 15 trials of different initial placements

Source 1	4	6	5	8	0	8	5	6	4	4	5	8	4	3	5
Source 2	14	0	15	0	12	0	15	5	8	3	7	0	16	7	3
Source 3	0	14	0	12	8	12	0	9	8	13	8	12	0	10	10
Sources captured	2	2	2	2	2	2	2	3	3	3	3	2	2	3	3

Fig. 5.6 Multi-robot system built for source localization experiments

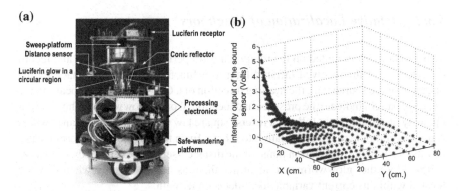

Fig. 5.7 **a** The wheeled mobile robot prototype, Glowworm, built to conduct the multiple sound source localization experiment. **b** Intensity distribution in the workspace due to the sound source

source localization experiments. The second version of Glowworms (Fig. 5.11), with modified hardware, were used for light source localization experiments.

5.3   Sound Source Localization

Each Glowworm is equipped with a sound pick up sensor in order to measure the intensity of sound at its location. A PIC16F877 microcontroller is used as the robot's processing unit. Next, we describe the Glowworm's hardware used to implement the modules of luciferin broadcast/reception and relative localization of neighbors.

5.3.1   Luciferin Glow/Reception

The glow consists of an infrared light modulated by an 8-bit serial binary signal that is proportional to the Glowworm's luciferin value at the current sensing-decision-action cycle. Four emitters that are mounted vertically and symmetrically about the Glowworm's central axis cast the infrared-light onto a buffed aluminium conic reflector (with an azimuth of 45°) in order to obtain a near circular pattern of luciferin emission in the Glowworm's neighborhood. Two infrared receivers mounted on a sweep-platform are used as luciferin receptors. In order to avoid problems due to interference between data signals coming from different neighbors, the receiver sweeps and aligns along the line-of-sight with a neighbor before reading the luciferin data transmitted by it. Using the above scheme, a minimum threshold separation of 10 cm. between neighbors was observed to be sufficient in order to distinguish data coming from different glowworm neighbors.

5.3.2   Relative Localization of Neighbors

Two photodiodes mounted on the rotary platform perform a 180° sweep and record the intensity of the received infrared light as a function of the angle made by the direction of the source with the heading direction of a Glowworm. Local peaks in the intensity pattern indicate the presence and location of other glowworms. The received intensity of infrared light follows a inverse square law with respect to distance which is used to compute range information to others. Even though the Glowworm locates all others within the perception range of the distance-sensor (excepting those that are eclipsed by other glowworms), it identifies them as neighbors only when they are located within its current variable local-decision domain.

5.3.3   Physical Experiments

Experiment 1: Sensing-decision-action Cycle

We describe a preliminary experiment that demonstrates the basic behavioral primitives used by a Glowworm to taxis toward source locations. A set of three Glowworms

A, B, and C are initially placed on the corners of an equilateral triangle of side 50 cm. Glowworms A and B remain stationary and emit a luciferin value of 128 and 60, respectively. Glowworm C emits a luciferin value of 40. At each iteration, the sweep platform homes by turning clock-wise until it makes a right angle with the heading direction of the Glowworm (Fig. 5.8b represents the end of homing phase at the first iteration), performs a 180° scan to acquire intensity samples and localize the neighbors, and aligns along the line-of-sight of each neighbor to receive the luciferin value emitted by it (Figs. 5.8c and d). The sensing phase of the first cycle is com-

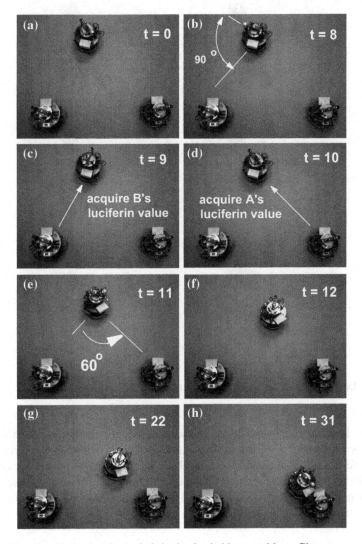

Fig. 5.8 Experiment showing the basic behavioral primitives used by a Glowworm to achieve taxis-behavior towards source locations

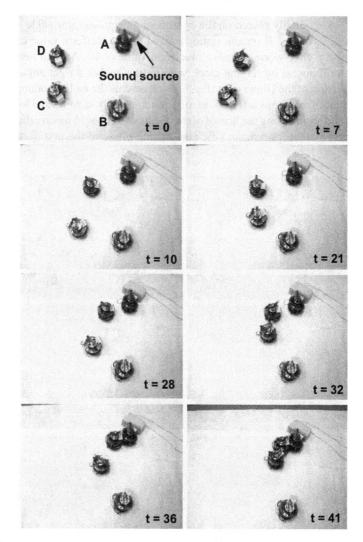

Fig. 5.9 Demonstration of a sound-source localization task

pleted at 10 s. For simplicity, we use a maximum-neighbor selection rule in which a Glowworm selects to move towards a neighbor that emits maximum luciferin. Since A's luciferin value is more than that of B, C decides to move towards A at every step. The first cycle is completed when the agent turns (Fig. 5.8e) and makes a step movement towards A at $t = 12$ s. Figure 5.8f C repeats the sensing-decision-action cycle twice before reaching A.

Experiment 2: Localization of a Single Sound Source

In this experiment, the glowworms localize a sound source, which is a loud speaker activated by a square wave signal of frequency 28 Hz. A microphone based sound sensor enables each Glowworm to measure the sound intensity at its current location. The sound-intensity pattern in the workspace, obtained by taking measurements at sufficiently large number of locations, is plotted in Fig. 5.7b. We place a Glowworm (A) near the source and a dummy Glowworm (B) away from the source which is kept stationary but made to emit luciferin proportional to the intensity measurement at its location. A is already located at the source and doesn't get a direction to move and hence, remains stationary. Initially, since B is in the vicinity of C (while A is not), it moves towards B. However, as it reaches closer to B it senses A and hence, changes direction in order to move towards A. Since D is closer to A, it makes deterministic movements towards A at every step. In this manner, the glowworms localize the sound source eventually. Snapshots from a video of the above experiment are shown in Fig. 5.9.

5.4   Light Source Localization

A PIC16F877 microcontroller is used as the robot's processing unit. The hardware architecture of second version of Glowworms is shown in Fig. 5.10. The various hardware modules of each robot that are used to implement the GSO algorithm for light localization are shown in Fig. 5.11. Experiments are conducted in which robots use GSO to localize a light source. A tethered power supply (refer to Fig. 5.6) was used in these experiments.

5.4.1   Luciferin Glow/Reception

As GSO is used in these experiments for locating a light source, a light pick-up sensor (M1) is used to measure the light intensity at the robot's location. The light sensor output is fed to an in-built analog-to-digital (A/D) convertor of the microcontroller. The output of the A/D module is converted into a luciferin glow that consists of an infrared emission modulated by an 8-bit serial binary signal that is proportional to the robot's luciferin value at the current sensing-decision-action cycle. Eight IR emitters (M4) that are mounted symmetrically about the robot's central axis are used to broadcast luciferin data in its neighborhood. An infrared receiver (M3) mounted on a sweep platform (M2) is used as a luciferin receptor. In order to avoid interference between data signals coming from different neighbors, the receiver sweeps and aligns along the line-of-sight with a neighbor to read the luciferin data broadcasted by it.

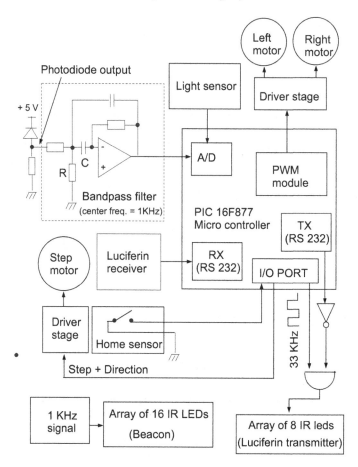

Fig. 5.10 The hardware architecture of Kinbots

5.4.2 Relative Localization of Neighbors

The various modules and sensors are marked as M1, M2, . . . , M8 in the Fig. 5.11.
An improved neighbor localization hardware module is built for the second version
of Glowworms: a circular array of sixteen infrared LEDs (M7 in Fig. 5.11), placed
radially outward, serves as a beacon to obtain a near circular emission pattern around
the robot. These IR LEDs are actuated by a 1 KHz square wave signal. A photodiode
(M6), mounted on a rotary platform (M2), performs a sweep and records the intensity
of the received infrared light as a function of the angle made by the direction of
the source with the heading direction of the robot. Local peaks in the intensity
pattern indicate the presence of neighbors and serve to compute their location in polar
coordinates. In particular, the received intensity of infrared light follows a inverse
square law with respect to distance, which is used to compute range information to

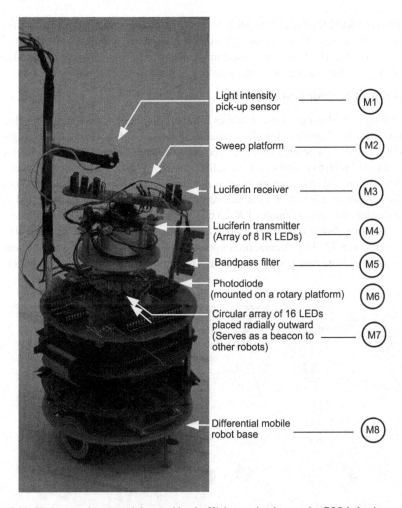

Light intensity pick-up sensor — (M1)

Sweep platform — (M2)

Luciferin receiver — (M3)

Luciferin transmitter (Array of 8 IR LEDs) — (M4)

Bandpass filter — (M5)

Photodiode (mounted on a rotary platform) (M6)

Circular array of 16 LEDs placed radially outward (Serves as a beacon to — (M7) other robots)

Differential mobile robot base — (M8)

Fig. 5.11 Various hardware modules used by the Kinbots to implement the GSO behaviors

robot-neighbors and the angular position of the local peak is approximated as the relative angle to a neighbor. The photo-diode output is passed through a bandpass filter, centered around a frequency of 1 KHz, in order to make it sensitive only to the frequency of the actuation signal of the infrared-beacons and reject noise due to ambient light conditions and IR signals used to broadcast luciferin information.

5.4.3   Physical Experiments

The obstacle avoidance maneuver is achieved by the movement of the robot along an arc in physical simulations. However, in real-robot implementation, when a robot approaches closer than a safe distance $d_{safe} < s$ to another robot, it performs a simple collision-avoidance maneuver through a discrete sequence of point-turn and straight-line movements. The resulting change in the obstacle avoidance behavior also leads to a change in the realization of the leapfrogging behavior of the robots.

Experiment 1: Basic GSO Behavior

According to GSO, a glowworm moves toward a neighbor whose luciferin value is relatively higher than that of itself. It remains stationary either when it has the highest luciferin value in its neighborhood or when it is isolated. The above basic behavior is implemented with two Kinbots in the following way. The first Kinbot (K_A), which is initially isolated, starts scanning its environment for neighbors using its sweep-sensor module. Another Kinbot (K_B) is allowed to move from a far away point and stop once it enters the sensing range of K_A. The robots emit constant luciferin values with $\ell_A < \ell_B$. Snapshots taken from a video of the above experiment at 2 s intervals are shown in Fig. 5.12. The circular arc imposed on the snapshots represents the sensor range of K_A. Note that it shifts its relative position from one snapshot to the other

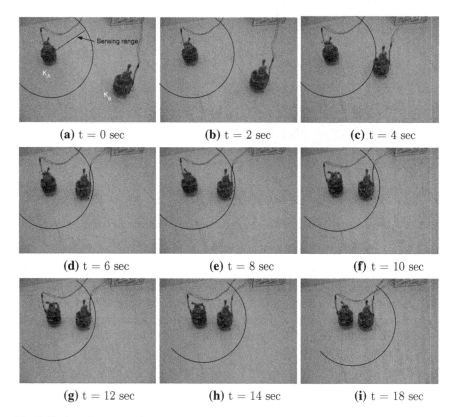

(a) t = 0 sec   **(b)** t = 2 sec   **(c)** t = 4 sec

(d) t = 6 sec   **(e)** t = 8 sec   **(f)** t = 10 sec

(g) t = 12 sec   **(h)** t = 14 sec   **(i)** t = 18 sec

Fig. 5.12 Snapshots taken from a video of the experiment used to demonstrate the basic GSO behavior

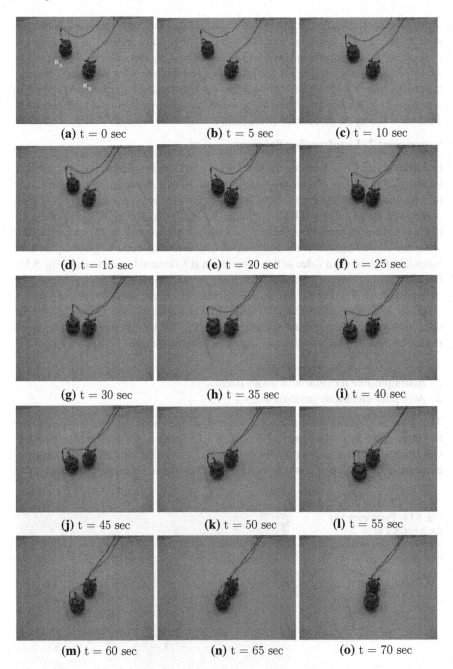

Fig. 5.13 Demonstration of the leapfrogging behavior

as K_A advances toward K_B. At $t = 0$ s, K_A is isolated and hence remains stationary until K_B enters its neighborhood at $t = 6$ s. At this point of time, K_A senses the presence of K_B, performs a point-turn until its heading direction is aligned along the line-of-sight with K_B, reads the luciferin value broadcast by K_B, starts moving toward K_B (since $\ell_A < \ell_B$) at $t = 8$ s, traverses a distance of about 10 cm, and stops moving at $t = 10$ s. Note that the sensing-decision-action cycle time is approximately 4 s. This 4 *sec*-cycle is repeated twice from $t = 10$ s to $t = 18$ s.

Experiment 2: Leapfrogging Behavior

Experiments with two Kinbots K_A and K_B are conducted to demonstrate the implicit leapfrogging behavior caused due to the incorporation of obstacle avoidance behavior into the basic GSO algorithm. Similar to the previous experiment, the robots emit constant luciferin values with $\ell_A < \ell_B$ and K_B is kept stationary. K_B is placed such that it is within the sensing range of K_A. However, the heading of K_A is, not aligned along the line-of-sight with K_B and instead, kept at a small acute angle away from it. Snapshots taken from a video of this experiment at 5 s intervals are shown in Fig. 5.13. Initially, K_A performs a point turn such that it is oriented toward K_B and then starts moving towards K_B (Fig. 5.13b). As K_A reaches closer than about 10 cm to K_B at $t = 20$ s (Fig. 5.13e), it toggles from the basic GSO behavior of moving toward a leader to an obstacle avoidance behavior at $t = 25$ s (Fig. 5.13f). The collision avoidance maneuver consists of the following sequence of movements:

1. A clock-wise point-turn of about 45°.
2. A straight-line movement of about 10 cm.
3. A counter clockwise point-turn of 45°.

The above sequence is repeated for another two times until the inter-robot distance becomes more than 10 cm at $t = 40$ s (Fig. 5.13i). Now, K_A switches to its basic GSO behavior and resumes to move toward K_B. Thereafter, this process of toggling between basic GSO and obstacle avoidance behaviors is repeated that gives rise to

Fig. 5.14 Paths traced by the robot K_A as it approaches, and leapfrogs over K_B

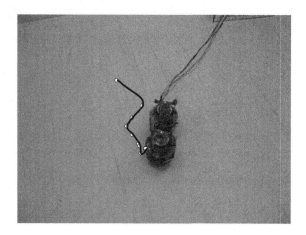

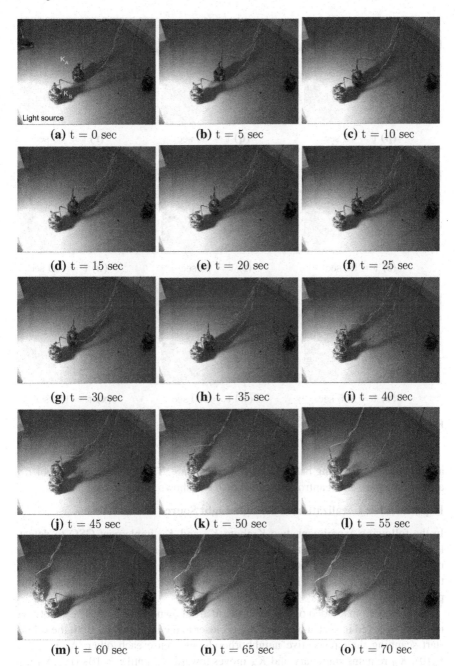

Fig. 5.15 Snapshots taken from a video of the light source localization experiment. Snapshots taken from a video of the light source localization experiment

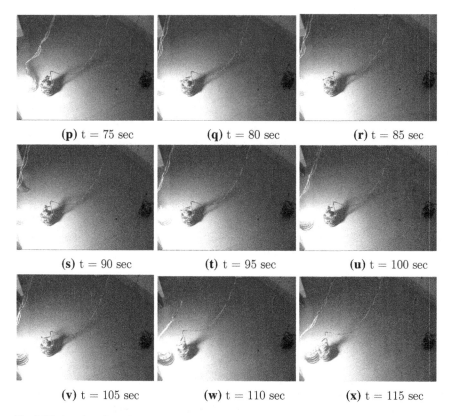

(**p**) t = 75 sec   (**q**) t = 80 sec   (**r**) t = 85 sec

(**s**) t = 90 sec   (**t**) t = 95 sec   (**u**) t = 100 sec

(**v**) t = 105 sec   (**w**) t = 110 sec   (**x**) t = 115 sec

Fig. 5.15 (continued)

an implicit leap-frogging behavior of K_A around K_B. The path traced by robot K_A as it approaches, and leapfrogs over, robot K_B is shown in Fig. 5.14.

Experiment 3: Localization of a Single Light Source

In the following experiment, two Kinbots K_A and K_B implement GSO to detect, taxis toward, and co-locate at a light source. The robots provide simple light intensity measurements at their current locations as inputs to the GSO algorithm running on their onboard processors. The robots use a photodiode based light sensor for this purpose.

Snapshots from a video of this experiment are shown in Fig. 5.15. The robots are initially deployed in such a way that K_B is closer to the light source. Both the robots start scanning their respective neighborhoods and sense each other. As $\ell_A(0) < \ell_B(0)$, K_B remains stationary and K_A moves toward K_B until $t = 10$ s (Fig. 5.15c). However, between $t = 10$ s and $t = 30$ s (Fig. 5.15g), K_A remains stationary. This can be attributed to the fact that the luciferin value received by K_A is corrupted. After $t = 30$ s K_A resumes to move toward K_B until $t = 35$ s (Fig. 5.15h) when it switches to obstacle avoidance behavior. Note that K_B starts moving toward K_A at t= 45 s

Fig. 5.16 Paths traced by
the robots K_A and K_B as
they taxis toward, and
co-locate at, the light source

Fig. 5.17 Toggling between
basic GSO and obstacle
avoidance behaviors of the
two robots K_A and K_B as
they interact with each other
and search for the light
source

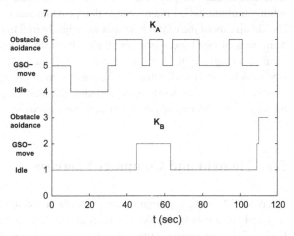

(Fig. 5.15j). However, at $t = 45$ s, K_A is still relatively far from the light source than
K_B. This can be attributed to the fact that when the robots are very close to each other,
the difference in luciferin values interferes with the range of sensor noise. K_B stops
moving at $t = 65$ s (Fig. 5.15n). However, at $t= 110$ s (Fig. 5.15w), K_A is closer to
the source and $\ell_{K_A}(110) > \ell_{K_B}(110)$. Therefore, K_B moves toward K_A. This results
in the robot pair moving further close to the light source.

The paths traced by the two robots as the two robots execute the modified GSO
algorithm is shown in Fig. 5.16. The toggling between the basic GSO and obstacle
avoidance behaviors of the two robots as they interact with each other, which eventually leads to their localization at the light source is shown in Fig. 5.17. Note from the
figure that the robots are idle during certain time intervals, which can be attributed
to one of the following reasons:

1. A robot is isolated.
2. A robot is measuring the highest light intensity.

3. Both the robots are measuring almost the same light intensity (A robot is made to move toward another one only when $|\ell_A - \ell_B| > 3$).
4. Reception of luciferin data is corrupted.

From Fig. 5.17, it is also clear that the sensing-decision-action cycles of the robots are asynchronous with respect to each other.

5.5 Summary

In this chapter, physical simulations and real-robot-experiments were conducted to test GSO's applicability to signal source localization tasks. The modifications incorporated into the algorithm in order to make it suitable for a robotic implementation were discussed. The physical robot simulations were carried out by using a multi-robot-simulator called Player/Stage that provides realistic sensor and actuator models. Details about the real robots that were hardwired to perform the basic behavioral primitives of GSO were presented. Results from experiments in which two robots localize a single light source were reported. By using more number of these robots, GSO can be used for effective localization of multiple signal sources. Till now, GSO was used for homogeneous agent swarms. The next chapter explores the idea of how the algorithm can be used in heterogeneous swarms.

5.6 Thought and Computer Exercises

Exercise 5.1 List the primary robotics applications of GSO. What modifications are required for a robotic implementation of GSO? Extend the MATLAB code of the basic GSO to incorporate these modifications. Assume that the robot

- has a circular foot print with a diameter of 10 cm
- uses differential steering
- can reach a linear speed of 0.5 m/s instantaneously
- can reach an angular speed of $\pi/2$ radians/s

Use the collision avoidance model given in Sect. 5.1.

Exercise 5.2 Use the code from above to simulate a group of five robots searching for a single signal source with the following distribution:

$$J_2(x, y) = \sum_{i=1}^{Q} a_i \exp(-b_i((x - x_i)^2 + (y - y_i)^2)) \tag{5.2}$$

where $Q = a_1 = b_1 = 1$ and $(x_1, y_1) = (0, 0)$. Choose a search space of $[-5$ m, 5 m$] \times [-5$ m, 5 m$]$ to deploy the robots. Conduct a set of 30 trials and report your observations. (Hint: The signal source may lie within the convex-hull of the initial deployment of the robots in some trials. This may not be the case in other trials. Do you notice any difference in group behaviors based on initial placements?) Report the average time taken by the robots to reach the source t_{travel}. What is the average computational time for convergence t_{comp}?

Exercise 5.3 Repeat the previous experiment to study how the number of robots impacts performance by varying n from 2 to 10 in increments of one and conducting the set of 30 trials for each case. Plot t_{travel} and t_{comp} as a function of n. Discuss your observations.

Exercise 5.4 Change the values of parameters in (5.2) such that the number of sources is increased to three and the sources are at a distance of 1 m from each other. Extend the experiments in Exercises 5.2 and 5.3 to study the influence of n on the number of sources captured.

Exercise 5.5 Port the MATLAB simulations into Player/Stage and repeat the experiments from Exercises 5.2, 5.3 and 5.4. What changes do you notice? For example, do you see any reduction in computational time? Do the trends in performance metrics like number of sources captured and time to reach sources look similar? What factors may cause differences, if any?

Exercise 5.6 Infrared based sensing hardware is used by the robotic glowworms described in this chapter to localize neighbors. Describe the method by which the distance and angle to all neighbors are determined by using this sensing scheme. What other sensing schemes can be used to achieve relative localization of neighbors in a robotic swarm?

Exercise 5.7 A circular array of infrared LEDs is used to emulate the luciferin glow in the robotic glowworms. Explain how the array is used to achieve the luciferin broadcast mechanism and how interference issues are resolved. How does a glowworm distinguish between the glows broadcasted from different neighbors? One-to-one communication model is another way by which glowworms can interact with each other. Compare the two approaches in terms of implementation complexity, scalability, and other advantages/disadvantages.

Exercise 5.8 The processing hardware is based on a PIC16F877 microcontroller. Design a Arduino based hardware architecture to achieve similar functionalities in the robots. Write a Arduino code to implement the GSO algorithm.

Chapter 6
Applications to Ubiquitous Computing Environments

In the previous chapter, physical simulations and real robot experiments were used to show the potential application of GSO to signal source localization tasks. The basic GSO algorithm, which was applied to multimodal optimization and signal source localization problems, considers a homogeneous swarm of mobile agents. In this chapter, a GSO variant for a heterogeneous swarm of mobile and stationary agents is developed and its application to ubiquitous computing environments is discussed. In particular, a GSO based ubiquitous computing environment is proposed to address the problem of sensing hazards. These hazards could manifest themselves in various forms such as nuclear leaks, health-risk causing gas or inflammable gas leakages in factories, fires caused by electrical short-circuits/sparks, etc.

The proposed ubiquitous computing environment comprises the deployment of a heterogeneous swarm of two types of nano-agents into the region of interest: (*i*) *Stationary nano-agents* that are embedded into the environment and act only as fixed sensors and beacons at locations of their deployment and (*ii*) *Mobile nano-agents* that use the information cues received from stationary and/or mobile neighbors in order to find their way to the source locations of the hazard. The agents deployed in the environment implement a modification to the GSO algorithm. This modified algorithm is unlike the previous GSO implementations where homogeneous mobile agent swarms are considered. The various phases of the GSO variant for heterogeneous swarms are described and some theoretical results related to the luciferin update mechanism are provided. Finally, simulation experiments are performed to demonstrate the efficacy of the algorithm in tackling such hazardous situations. It is shown that the deployment of stationary agents in a grid-configuration leads to multiple phase-transitions in a graph of minimum number of mobile agents required for 100% source-capture versus the number of stationary agents.

© Springer International Publishing AG 2017   157
K.N. Kaipa and D. Ghose, *Glowworm Swarm Optimization*,
Studies in Computational Intelligence 698, DOI 10.1007/978-3-319-51595-3_6

6.1   Ubiquitous Computing Based Environments

Ubiquitous computing based environments may be defined as human surroundings that are furnished with a network of intelligent computing devices, which could be either stationary or mobile (or an assortment of both), in order to service certain tasks. Mark Weiser introduced ubiquitous computing, in its current form, in 1988 and wrote some of the earliest papers on ubiquitous computing [216]. Ubiquitous computing based environments have several applications to the industry like environmental monitoring [101], ubiquitous factory environments [87], and self-sensing spaces [52]. Kim et al. [101] develop a framework that uses ubiquitous sensor networks for atmospheric environment monitoring. Jabbar et al. [87] present methods that integrate latest technologies like RFID, PDA, and Wi-Fi in order to transform a nuclear power plant into an ubiquitous factory environment where performance benefits such as effective data communication between local area operators and control room and minimization of work duration and errors in the wake of safety requirements are achieved. El-Zabadani et al. [52] propose a novel approach to mapping and sensing smart spaces in which a mobile platform equipped with on-board RFID modules identifies and locates RFID-tags that are embedded in the carpet in the form of a grid.

An important application of particular interest is that of timely hazard sensing in an environment where the undetected hazard condition can snowball into a major disaster. Several examples of such scenarios exist today. One such example is a nuclear installation where the slightest leakage can signal a major potential disaster unless the leak is quickly controlled. Other similar scenarios in a factory environment are health-risk causing gas or inflammable gas leakage in factories, uncontrolled and unwanted vibrations or other mechanical phenomena leading to gradual failure of critical structures, etc. Even at homes, a large number of fire calamities are caused by electrical discharges of malfunctioning wire joints.

6.2   Heterogeneous Agent Swarms

An ubiquitous computing environment can be envisaged to have an assortment of a large group of virtually invisible nano-agents that can be used to detect potential hazard situations and, if possible, neutralize them. This group is composed of two types of agents:

1. *Stationary nano-agents* that are embedded into the environment and act only as fixed beacons at locations of their deployment.
2. *Mobile nano-agents* that use the information cues received from stationary and/or mobile neighbors in order to detect, taxis toward, and gather at potential hazard locations (thereby making themselves visible to an external sensor) in order to prevent an impending disaster.

Moreover, if there are multiple sources of hazard then this group of mobile nano-agents should be able to sense, automatically split into sub-groups, and simultaneously home on to these individual sources. The advantages of using an assortment of stationary and mobile agents, instead of using a homogeneous mobile swarm, are two fold:

1. Since the stationary sensors could be built at very low costs, they can be deployed in large numbers. This results in a need for using only a few number of mobile agents to accomplish a hazard sensing task that would otherwise require a relatively large number of mobile agents for its completion.
2. Even though the mobile agents may be deployed at locations that are far away from the sources (e.g., corners of the region), as the stationary beacons pervade the entire environment and provide information about the signal profile at the locations of their deployment, the mobile agents use such information to find their way to a hazard source and, once in the vicinity of the source, the mobile agents exchange information with each other in order to home on to the source location.

6.2.1 GSO Variant for Heterogeneous Agent Swarms

The GSO variant starts with the placement of the stationary and mobile agents randomly in the environment so that they are well dispersed. Let S_{agents} and M_{agents} represent the sets of all stationary and mobile agents, respectively. Let $S_a = |S_{agents}|$ and $M_a = |M_{agents}|$ be the number of stationary and mobile agents, respectively. We define the *assortment ratio* a_r as the ratio of number of mobile agents M_a to the total number of agents n:

$$a_r = \frac{M_a}{M_a + S_a}; n = M_a + S_a \qquad (6.1)$$

Initially, all the agents contain an equal quantity of luciferin ℓ_0. Each iteration consists of a luciferin-update phase followed by a movement-phase based on a transition rule.

Luciferin-update phase: The luciferin update rule for a mobile agent is identical to the update rule (2.1) given in the basic GSO algorithm (refer to Chap. 2).

However, the luciferin update rule for a stationary agent $j \in S_{agents}$ is given by:

$$\ell_j(t + 1) = (1 - \rho)\ell_j(t) + \gamma J(x_j) \qquad (6.2)$$

where, $J(x_j)$ represents the value of the signal-profile at the stationary agent j's location x_j, which remains fixed for all time t.

Movement-phase: A mobile agent does not distinguish between mobile and stationary neighbors. Accordingly, during the movement-phase, each mobile agent decides, using a probabilistic mechanism, to move toward a neighbor (stationary or mobile)

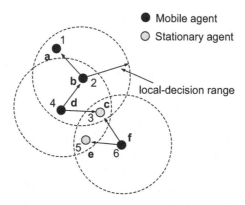

Fig. 6.1 Emergence of a directed graph based on whether the agent is stationary or mobile, the relative luciferin level of each agent, and availability of only local information. Agents are ranked according to the increasing order of their luciferin values. For instance, the Agent a whose luciferin value is highest is ranked '1' in the figure

that has a luciferin value more than its own. Note that the stationary agents do not execute this phase. Figure 6.1 shows the emergence of a directed graph among a set of six agents based on whether the agent is stationary or mobile, their relative luciferin levels, and availability of only local information. In the figure, a, b, d, and f are mobile agents where c and e are stationary agents. Since agent a has the highest luciferin among all the agents, it remains stationary. Mobile agent b has only one direction of movement whereas agent d could move toward either b or c. Similarly, agent f has two possible directions of movement. For each mobile agent i, the probability of moving toward a neighbor j is given by:

$$p_j(t) = \frac{\ell_j(t) - \ell_i(t)}{\sum_{k \in N_i(t)} \ell_k(t) - \ell_i(t)} \tag{6.3}$$

where, $j \in N_i(t) \neq \emptyset$, and

$$N_i(t) = \{j : j \in S_{agents} \bigcup M_{agents} \text{ and } d_{i,j}(t) < r_d^i(t) \text{ and } \ell_i(t) < \ell_j(t)\}$$

is the neighborhood of agent i at time t, $d_{i,j}(t)$ represents the Euclidean distance between agents i and j at time t, $\ell_j(t)$ represents the luciferin level associated with agent j at time t, $r_d^i(t)$ represents the variable local-decision range associated with agent i at time t, and r_s represents the radial range of the luciferin sensor. Let the agent i select an agent $j \in N_i(t)$ with $p_j(t)$ given by (6.3). Then the discrete-time model of the agent movements is given by (2.3) and is restated here:

$$x_i(t+1) = x_i(t) + s\left(\frac{x_j(t) - x_i(t)}{\|x_j(t) - x_i(t)\|}\right) \tag{6.4}$$

The local-decision domain range update rule for each mobile agent is given by (2.4) and is restated here:

$$r_d^i(t+1) = \min\{r_s, \max\{0, r_d^i(t) + \beta(n_t - |N_i(t)|)\}\} \tag{6.5}$$

6.2.2 Theoretical Expressions for Agents' Luciferin Levels

In Chap. 2, the two theoretical results provided for the case of homogeneous agents also hold for the mobile agents in the heterogeneous swarm. Here, the corresponding theoretical results for the case of heterogeneous agents are provided. In particular, expressions are provided for the luciferin levels at any time t for stationary and mobile agent cases, respectively. For stationary agents, the respective result is extended to find their luciferin levels at steady state.

Theorem 6.1 *For each stationary agent $i \in S_{agents}$, assuming that the update rule (6.2) is used, the luciferin value $\ell_i(t)$ at any time t is given by:*

$$\ell_i(t) = (1-\rho)^t \ell_0 + \left\{\frac{1-(1-\rho)^{t-1}}{\rho}\right\} \gamma J(x_i) \tag{6.6}$$

where, ℓ_0 is the initial luciferin level of agent i. Moreover, $\ell_i(t)$ asymptotically converges to $\left(\frac{\gamma}{\rho}\right) J(x_i)$.

Proof Given that the initial luciferin value of the stationary agent i is ℓ_0, its luciferin level at Iteration 1, using the luciferin update rule (6.2), is given by:

$$\ell_i(1) = (1-\rho)\ell_0 + \gamma J(x_i)$$

At Iteration 2:

$$\ell_i(2) = (1-\rho)^2 \ell_0 + [(1-\rho) + 1]\gamma J(x_i)$$

Proceeding in a similar way, at iteration t:

$$\ell_i(t) = (1-\rho)^t \ell_0 + [1 + (1-\rho) + (1-\rho)^2 + \cdots + (1-\rho)^{t-1}]\gamma J(x_i)$$
$$= (1-\rho)^t \ell_0 + \left[\frac{1-(1-\rho)^{t-1}}{\rho}\right] \gamma J(x_i) \tag{6.7}$$

Since $0 < \rho < 1$, it is evident from (6.7) that as $t \to \infty$, $\ell_i(t) \to \left(\frac{\gamma}{\rho}\right) J(x_i)$. $\qquad \square$

Theorem 6.2 *For each mobile agent $i \in M_{agents}$, assume that the update rule (2.1) is used and define $P(t)$ as $P(t) = (1-\rho)^t$. Now, the luciferin value $\ell_i(t)$, at any time t, is given by:*

$$\ell_i(t) = P(t)\ell_0 + \gamma \sum_{k=0}^{t-1} P(k)J(x_i(t-k)) \tag{6.8}$$

where, ℓ_0 is the initial luciferin level of agent i.

Proof Given that the initial luciferin value of the mobile agent i is ℓ_0, its luciferin level at Iteration 1, using the update rule (2.1), is given by:

$$\ell_i(1) = (1 - \rho)\ell_0 + \gamma J(x_i(1))$$

At Iteration 2:

$$\ell_i(2) = (1 - \rho)^2 \ell_0 + (1 - \rho)\gamma J(x_i(1)) + \gamma J(x_i(2))$$

Proceeding in a similar way, at iteration t:

$$\begin{aligned}
\ell(t) &= (1 - \rho)^t + (1 - \rho)^{t-1}\gamma J(x_i(1)) + (1 - \rho)^{t-2}\gamma J(x_i(2)) \\
&\quad + \cdots + (1 - \rho)\gamma J(x_i(t-1)) + \gamma J(x_i(t)) \\
&= (1 - \rho)^t + \gamma \sum_{k=0}^{t-1} (1 - \rho)^k J(x_i(t-k)) \tag{6.9}
\end{aligned}$$

Using the definition $P(t) = (1 - \rho)^t$, we simplify (6.9) as below:

$$\ell_i(t) = P(t)\ell_0 + \gamma \sum_{k=0}^{t-1} P(k)J(x_i(t-k)) \tag{6.10}$$

Thus we prove the result in (6.8). Note that the second term in the above equation is a discrete-convolution of $\gamma P(t)$ and $J(x_i(t))$ between the summation limits 0 and $t - 1$. □

6.3 Simulation Experiments

An environment where multiple sources of hazard, representative of examples discussed earlier, radiate signals that pervade the environment is considered. For this purpose, the $J_1(x, y)$ function (2.24) is used to create the signal profile. A search area of 36 sq. units ($[-3, 3] \times [-3, 3]$) is considered. The above signal-profile has peaks of different values (representing the signal level) at three different locations. It is assumed that propagation of the signal is instantaneous, which means any decay or enhancement of signals is immediately felt in the same way in the environment and the qualitative nature of the signal profile remains unchanged. The algorithm para-

Table 6.1 Values of algorithm parameters used in the simulations

ρ	γ	s	β	n_t
0.4	0.6	0.03	0.08	5

meters used in the simulations are shown in Table 6.1. Four experimental scenarios are considered that are described below.

6.3.1 Simulation Experiment 1: Effect of Assortment Ratio on Number of Captured Signal Sources

The goal of the first set of experiments is to study the influence of assortment ratio a_r as defined in (6.1), on the number of signal sources detected. Figure 6.2a–c show the emergence of movements of the mobile agents and their eventual co-location at the hazard source(s) for assortment ratios of 0.2, 0.3, and 0.4, respectively, when the total number of 100 agents was considered. Note that the number of sources detected increases with the increase in the assortment ratio for a fixed population size. In the third case (Fig. 6.2c), all the sources are detected within 200 iterations corresponding to a computation time of 2.67 s.[1] However, the number of iterations and simulation time give only a qualitative idea and the time taken for the movements of mobile agents have to be considered in order to compute the time spent by the swarm to complete the task in a realistic hazard sensing application. Moreover, the algorithm would run parallel in all the agents and each agent needs to perform only its own computations (as opposed to the computer simulation that is implemented in a sequential manner), thereby reducing the overall computation time.

Figure 6.3 shows the number of signal sources detected as a function of a_r for different population sizes. It is observed that the assortment ratio required for 100% source-capture decreases with increase in the total number of agents. Note that the number of mobile agents does not reduce in the same proportion as the assortment ratio. For instance, complete source-capture requires about 40–45 mobile agents for all three cases of population sizes ($n = 50, 100, 200$). However, it is shown in the next set of experiments that when the environment is densely populated with stationary agents, the minimum number of agents required for 100% source-capture begins to show a reducing trend with increase in the population size.

---

[1] The algorithm is coded using Matlab 7.0 and all simulations are implemented on a 2 GB RAM, 3 GHz Intel P4 machine with a Windows XP operating system.

Fig. 6.2 Emergence of the
movements of the mobile
agents and their eventual
co-location at the hazard
sources for different
assortment ratios a_r: **a**
$n = 100, a_r = 0.2.$ **b**
$n = 100, a_r = 0.3.$ **c**
$n = 100, a_r = 0.4$

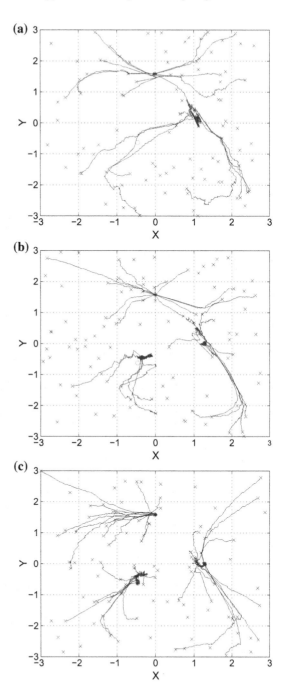

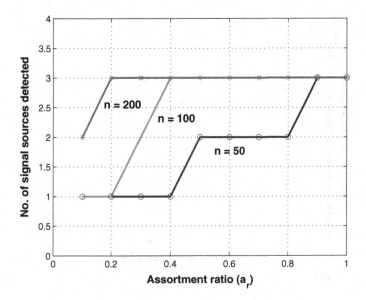

Fig. 6.3 Number of signal sources detected as a function of assortment ratio a_r for $n = 50$, 100, and 200

6.3.2 Simulation Experiment 2: Assortment Ratios for Capturing all Sources

The next set of experiments are conducted for different population sizes and assortment ratios corresponding to environments that are densely populated with stationary agents and sparingly populated with mobile agents. The basic motivation for this experimental scenario lies in the fact that the stationary sensors could be built at considerably low costs and hence they can be deployed in very large numbers. Consequently, the minimum number of mobile agents required to detect all the signal sources M_a^{min} is plotted as a function of the population size (Fig. 6.4a). Note that the change in the slope of M_a^{min} after $n = 400$ indicates a steeper decline in the minimum number of mobile agents for 100% source-capture with increase in population size. Figure 6.4b shows the emergence of movements of the mobile agents and their eventual co-location at the hazard sources when $n = 800$ and the number of mobile agents is only 15 ($a_r = 0.0187$). It can be inferred from this result and Fig. 6.4a that by using a large number of stationary sensors, only a few number of mobile agents is sufficient to accomplish a hazard sensing task that would otherwise require a relatively large number of mobile agents for its completion.

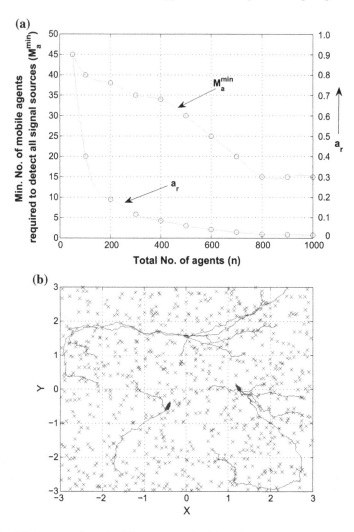

Fig. 6.4 **a** Minimum number of mobile agents required for 100% source-capture as a function of the total number of agents. **b** Emergence of the movements of the mobile agents and their eventual co-location at the hazard sources for $n = 800$ and $a_r = 0.0187$

6.3.3   Simulation Experiment 3: Pre-defined Deployment of Mobile Agents

Heterogeneous swarms are considered where stationary agents pervade the entire environment while the initial placement of the mobile agents is restricted to regions such as corners and edges of the environment. The results are compared with that of homogeneous swarms in order to demonstrate the efficacy of using an assortment of stationary and mobile agents in these placement scenarios. In particular, the following three deployments are considered:

Left-bottom corner region: Initially, a homogeneous group of five mobile agents are deployed in the region $[-3, -2] \times [-3, -2]$. (Experiments show that a minimum number of five agents is found to be necessary for a homogeneous mobile swarm in order to perform a robust search or foraging in the environment.) Figure 6.5a shows the emergence of agent-movements and co-location of the agents at a signal-source in about 600 iterations. However, when a heterogeneous swarm of 800 agents is considered, where a set of 795 stationary agents is deployed in the entire environment and the 5 mobile agents are deployed in the region $[-3, -2] \times [-3, -2]$, the mobile agents take only 100 iterations to capture the same source (Fig. 6.5b).

Right-edge region: In both the homogeneous and heterogeneous cases, a set of 20 mobile agents is deployed in the region $[2, 3] \times [-3, 3]$, which represents the right-edge region of the environment. In the heterogeneous agents case, an additional set of 780 stationary agents is deployed in the entire environment. Figure 6.6a and b show the emergence of mobile agent movements in the homogeneous and heterogeneous cases, respectively. The mobile agents are able to capture only one signal-source in the absence of the stationary agents, which can be explained in the following way: Note from Fig. 6.6a that the direction of swarm-advancement is heavily influenced by the initial location of a mobile agent that is closest to a nearby source, which does not move until another agent (with a relatively less luciferin value) reaches it and performs a leapfrogging move (described in Chap. 2) in order to over-take it. Consequently, the property of the leapfrogging mechanism enforces all the swarm members to move in a single group and eventually home on to a single signal-source. However, when stationary agents pervade the entire environment, they give rise to attraction-beacons at different locations corresponding to multiple sources, thereby enabling the exploration of the mobile-swarm in different directions in order to capture multiple sources. In Fig. 6.6b, the mobile-swarm splits into two subgroups and eventually captures two of the three signal-sources.

Four Corners: In the first set of experiments, groups of 6 and 3 mobile agents at each corner are deployed to constitute homogeneous swarms. Emergence of agent-movements within 500 iterations for each case is shown in Figs. 6.7a and 6.8a, respectively. In the next set of experiments, heterogeneous groups of 800 agents are considered where 6, 3, 2, and 1 mobile agent(s), respectively, are deployed at each corner and the rest of them, which are stationary agents, pervade the entire environment. Figures 6.7b, 6.8b, 6.9a, b show the emergence of mobile agent movements within 200 iterations for each case. Note from Fig. 6.9b that even a single mobile agent from a corner location could find its way to a nearby source unlike in the homogeneous agent case where a minimum number of five agents is required for the swarm to acquire a search capability.

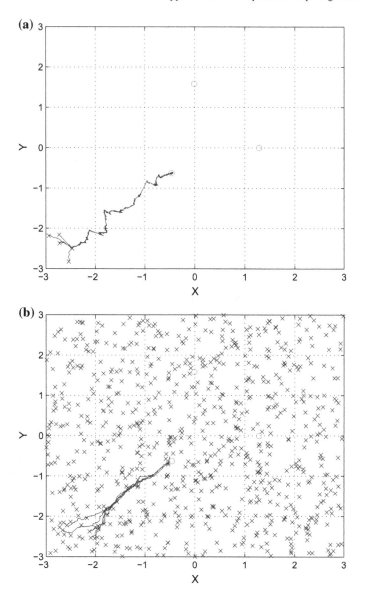

Fig. 6.5 a Homogeneous group of 5 mobile agents is deployed in the *left-bottom corner* of the environment. **b** Heterogeneous group of 800 agents is deployed with 5 mobile agents deployed in the *left-bottom corner* of the environment

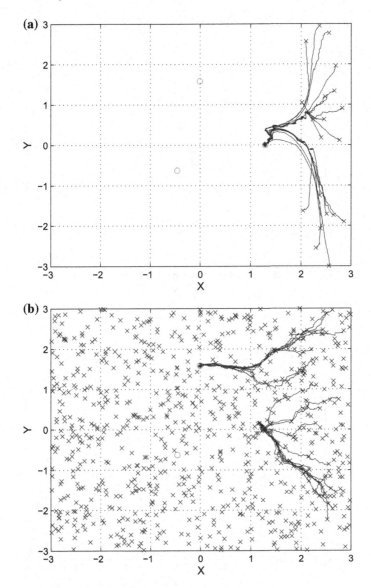

Fig. 6.6 **a** Homogeneous group of 20 mobile agents is deployed in the *right-edge* region of the environment. **b** Heterogeneous group of 800 agents is deployed with 20 mobile agents deployed in the *right-edge* region of the environment

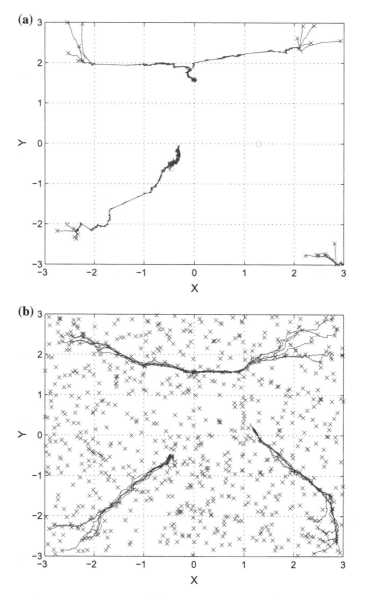

Fig. 6.7 **a** Homogeneous group of 6 mobile agents is deployed in each corner of the environment. **b** Heterogeneous group of 800 agents is deployed with 6 mobile agents in each corner of the environment

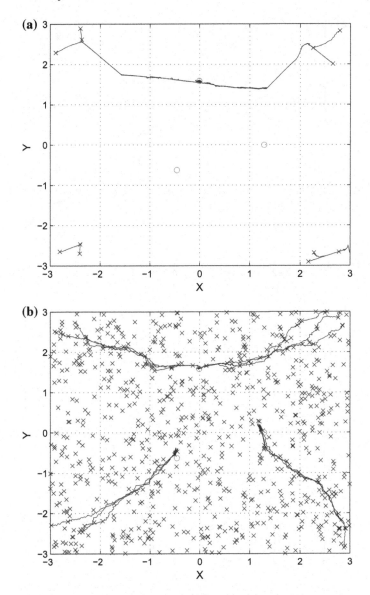

Fig. 6.8 **a** Homogeneous group of 3 mobile agents is deployed in each corner of the environment. **b** Heterogeneous group of 800 agents is deployed with 3 mobile agents in each corner of the environment

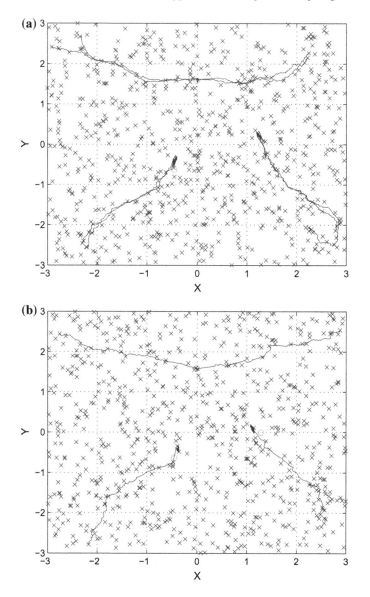

Fig. 6.9 **a** Heterogeneous group of 800 agents is deployed with 2 mobile agents deployed in each corner of the environment. **b** Heterogeneous group of 800 agents is deployed with 1 mobile agent in each corner of the environment

Fig. 6.10 Stationary agents deployed in a grid-configuration. The number of stationary agents that are within the sensing range of a mobile agent is a function of the spacing between the grid points

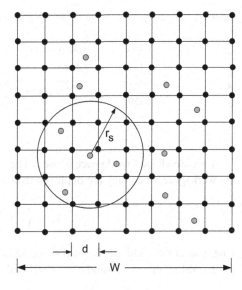

● - Stationary agent

○ - Mobile agent

Fig. 6.11 The signal profile used to examine algorithm's behavior when stationary agents are deployed in a grid-configuration

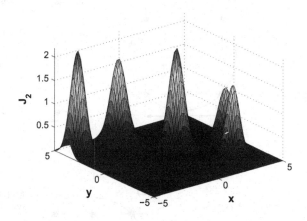

6.3.4 Simulation Experiment 4: Deployment of Stationary Agents in a Uniform Grid

In this set of experiments, the algorithm's behavior when the stationary agents are deployed in a grid-configuration, as shown in Fig. 6.10, is examined. The functions $J_1(x, y)$ (2.24) and $J_2(x, y)$ (3.1) are used for these experiments. A set of five peaks ($Q = 5$) is considered for the purpose (Fig. 6.11). The corresponding values of $\{a_i, b_i, x_i, y_i, i = 1, \ldots, 5\}$ used to generate the function $J_2(x, y)$ are shown in Table 6.2.

Table 6.2 Values of a_i, b_i, x_i, and y_i used to generate the function profile $J_2(x, y)$

	1	2	3	4	5
a_i	1.7206	2.0970	1.5235	2.1947	1.0986
b_i	2.1421	2.4017	2.9246	2.5010	2.4800
x_i	−0.6813	1.3427	3.0303	−4.1612	4.4546
y_i	4.1594	1.0199	−2.4644	3.7345	0.1340

A square-shaped workspace is used. The total number of stationary agents S_a in the workspace when they are deployed in a grid-configuration is given by:

$$S_a = \left(1 + \frac{w}{d}\right)^2 \tag{6.11}$$

where, w is the width of the workspace and d is the spacing between adjacent grid points. Values of $w = 10$ and $r_s = 1$ are used. Note from (6.11) and Fig. 6.10 that for

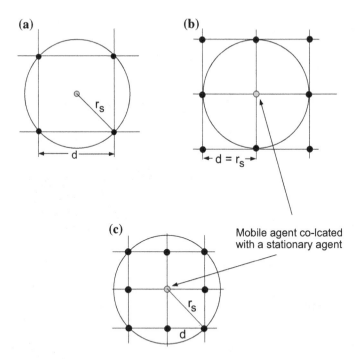

Fig. 6.12 The nature of the local neighborhood of a mobile agent for three different values of d. **a** When $d = \sqrt{2}r_s$, a mobile agent at the center of a grid-cell acquires four stationary agents as neighbors. **b** When $d = r_s$, a mobile agent that is co-located with a stationary agent acquires four stationary agents as neighbors. **c** When $d = \frac{r_s}{\sqrt{2}}$, a mobile agent that is co-located with a stationary agent acquires four additional stationary agents as neighbors (that is, totally eight neighbors)

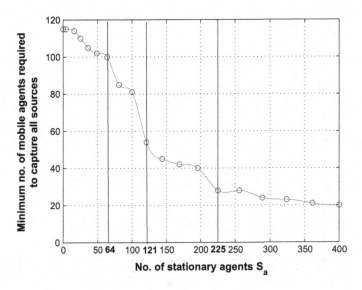

Fig. 6.13 The minimum number of mobile agents M_a^{min} for 100% source-capture as a function of the number of stationary agents S_a, when the signal-profile $J_1(x, y)$ is used

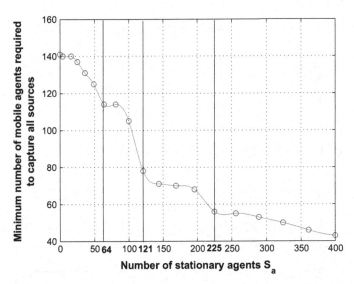

Fig. 6.14 The minimum number of mobile agents M_a^{min} for 100% source-capture as a function of the number of stationary agents S_a, when the signal-profile $J_2(x, y)$ is used

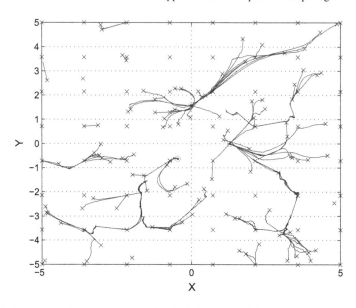

Fig. 6.15 Emergence of agent movements for the following conditions: Function $J_1(x, y)$ is used, $S_a = 64$, $M_a = 100$

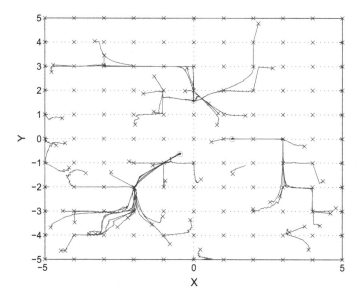

Fig. 6.16 Emergence of agent movements for the following conditions: Function $J_1(x, y)$ is used, $S_a = 121$, $M_a = 54$

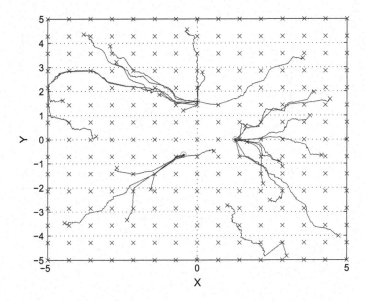

Fig. 6.17 Emergence of agent movements for the following conditions: Function $J_1(x, y)$ is used, $S_a = 225$, $M_a = 28$

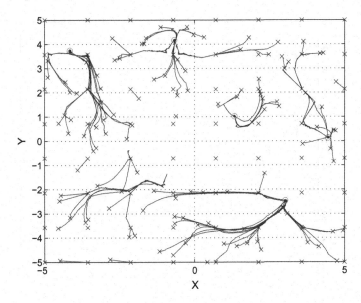

Fig. 6.18 Emergence of agent movements for the following conditions: Function $J_2(x, y)$ is used, $S_a = 64$, $M_a = 114$

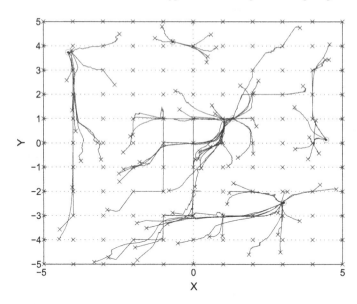

Fig. 6.19 Emergence of agent movements for the following conditions: Function $J_2(x, y)$ is used, $S_a = 121$, $M_a = 78$

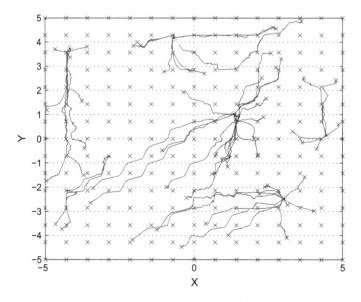

Fig. 6.20 Emergence of agent movements for the following conditions: Function $J_2(x, y)$ is used, $S_a = 225$, $M_a = 56$

fixed values of w and r_s, the number of stationary-agent-neighbors of a mobile agent generally increases with a decrease in the value of d. However, when d approaches the values of $\sqrt{2}r_s$, r_s, and $r_s/\sqrt{2}$, there is a leap-increment in the number of neighbors of a mobile agent. For instance, when $d > \sqrt{2}r_s$, a mobile agent that is located at the center of a grid-cell can only have other mobile agents as possible neighbors. However, when $d = \sqrt{2}r_s$, it acquires a set of four stationary agents as neighbors (Fig. 6.12a). Similarly, a mobile agent co-located at a stationary agent's location acquires four stationary-agent-neighbors when $d = r_s$ (Fig. 6.12b). It acquires an additional set of four neighbors when $d = r_s/\sqrt{2}$ (Fig. 6.12c). The influence of the spacing d on the number of neighbors of a mobile agent, as described above, leads to multiple phase-transitions in the heterogeneous-swarm behavior, which can be captured when the minimum number of mobile agents required to capture all the sources is plotted as a function of the number of stationary agents.

In each experiment, the number of stationary agents used is kept as a perfect-square in order to ensure a symmetric deployment of the stationary agents. The values of S_a that are perfect-squares and corresponding to values of d that are nearest to $d = \sqrt{2}$, 1, and $1/\sqrt{2}$ (since $r_s = 1$) are 64, 121, and 225, respectively. Figures 6.13 and 6.14 show the plots of the minimum number of mobile agents M_a^{min} required to capture all the sources as a function of S_a when the functions $J_1(x, y)$ and $J_2(x, y)$ are used, respectively. At each value of S_a, a set of ten experimental trials is conducted and M_a^{min} is chosen as the minimum number of mobile agents for which 100% source-capture is achieved in all the ten trials. Note that, from Figs. 6.13 and 6.14, the occurrence of phase-transitions, in the neighborhoods of $S_a = 64$, 121, and 225, shown in simulations are in perfect agreement with the analytical predictions made above. Moreover, the similarities between plots in Figs. 6.13 and 6.14 indicate the fact that phase transition behavior of the heterogeneous-swarm is invariant to the kind of multimodal signal profile spread in the environment. A minor difference between the two plots is a vertical shift of the graph in Fig. 6.14, which can be attributed to the fact that the function $J_2(x, y)$ has more peaks than the function $J_1(x, y)$ and hence requires relatively more number of mobile agents for 100% peak-capture. The emergence of agent-movements and eventual capture of all the sources at various phase-transition points are shown in Figs. 6.15, 6.16, 6.17, 6.18, 6.19 and 6.20. Note from Figs. 6.16 and 6.19 that some of the agents move on the edges of the grid-cells. This occurs due to the fact that in these cases, since $d = r_s$, an agent that is co-located with a stationary agent has stationary-agent-neighbors on the edges of four neighboring grid-cells as shown in Fig. 6.12b and hence moves toward one of them if there are no other mobile agent-neighbors.

6.4 Summary

A ubiquitous computing environment comprising a heterogeneous swarm of nano-agents using GSO was proposed for hazard sensing applications. The agents implement a modification to GSO in order to sense and eventually co-locate at the source locations of the hazard. The heterogeneous swarm is composed of an assortment of stationary and mobile agents where the stationary agents are embedded into the environment and act only as stationary sensors, while the mobile agents use the information cues received from stationary and/or mobile neighbors in order to find their way to the source locations. Simulations in various experimental scenarios and comparison of results with that of homogeneous swarms demonstrate the significance of supplementing a mobile agent group with a large number of inexpensive stationary agents. It was also shown that a grid based deployment of stationary agents leads to multiple phase-transitions in the heterogeneous swarm behavior. In the next chapter, the behavior of the agents in GSO when the source locations are allowed to move is examined and the application of GSO to pursuit of mobile signal sources is discussed.

6.5 Thought and Computer Exercises

Exercise 6.1 Define ubiquitous environments. Enumerate the primary applications of ubiquitous environments. Among these, what are the applications that can be designed in the framework of GSO? Provide a detailed rationale as to why GSO is suitable to these applications. What modifications to basic GSO are needed for application to this class of problems?

Exercise 6.2 The GSO variant discussed in this chapter involves a heterogeneous swarm of mobile and stationary glowworms. How does heterogeneity affect the functionality of GSO? How does it affect the connectivity in the swarm?

Exercise 6.3 Consider the Rastrigin's function (4.1) in two dimensions ($m = 2$) in a search space of $[-3, 3] \times [-3, 3]$:

$$J(x, y) = 20 + x^2 + y^2 - 10\cos(2\pi x) - 10\cos(2\pi y)$$

1. Write a MATLAB code to implement the heterogeneous variant of GSO on this function.
2. The assortment ratio a_r was defined in this chapter as the number of mobile agents M_a to the total number of mobile agents and stationary agents S_a. Run simulations to conduct 30 trials each and report the average number of peaks obtained for different combinations of M_a and S_a. In particular, consider the following cases:

   (a) $a_r = 1$ ($S_a = 0$; reduces to a baseline case of a homogeneous swarm). Find the minimum value of M_a for which all the peaks are captured.
   (b) $0 < a_r < 1$. Start with the value of M_a found in the baseline case and $S_a = M_a$ and ensure that all peaks are captured. Next, keep decreasing M_a and record the the number of peaks captured. What is the new value for the minimum M_a needed to capture all the peaks for the chosen S_a? What is the corresponding value of a_r?

(c) Keep incrementing S_a until a handful of agents ($\approx 2\times$ Number of peaks) is sufficient to capture all the peaks. What are the new values of S_a and a_r?

3. Based on the above experiments, comment on the the influence of assortment ratio on swarm connectivity and number of sources captured.

Exercise 6.4 Consider the test function given in the previous exercise. Now, consider a predefined deployment of mobile agents. In particular, consider the following deployment cases:

1. Consider a small square-region at each corner of the workspace and randomly deploy an equal number of mobile agents in each square.
2. Consider a small rectangular-region at each edge of the workspace and randomly deploy an equal number of mobile agents in each rectangle.

Conduct similar experiments to characterize the influence of assortment ratio on the number of sources captured. Comment on your observations.

Exercise 6.5 Consider the test function given in Exercise 6.3. Now, consider a grid placement of stationary agents, where the number of stationary agents $S_a(d)$ is a function of adjacent distance between grid points, for a fixed workspace of width w. Conduct experiments to obtain a graph of minimum number of mobile agents required for 100% peak-capture as a function of $S_a(d)$. Provide a rationale for the occurrence of any phase transitions that you may observe in the graph (Hint: Using the relationship between d and sensing range r_s and the impact it has on neighbor connectivity).

Chapter 7
Pursuit of Multiple Mobile Signal Sources

The previous chapter dealt with the application of a GSO variant to ubiquitous computing environments. Stationary sources were considered in all the previous chapters. However, in this chapter, the behavior of GSO agents in the presence of mobile sources is investigated. In particular, a coordination scheme based on GSO is developed that enables a swarm of glowworms to pursue a group of mobile signal sources. Examples of such sources include hostile mobile targets in a battlefield and moving fire-fronts that are created in forest fires. We assume that the mobile source radiates a signal whose intensity peaks at the source location and decreases monotonically with distance from the source.

For the case where the positions of the pursuers and the moving source are collinear, a theoretical result that provides an upper bound on the relative speed of the mobile source below which the pursuers succeed in chasing the source is presented. Simulations are used to demonstrate the efficacy of the algorithm in addressing these pursuit problems. In particular, simulation results for single and two source cases are presented, respectively; each source is assumed to move in a circular trajectory and at constant angular speed. In the case where the positions of the pursuers and the moving source are non-collinear, numerical experiments are used to determine an upper bound on the relative speed of the mobile source below which the pursuers succeed in chasing the source.

7.1 Group Hunting in Natural World

Principles from the area of swarm intelligence [18], in particular, animal behavior [59, 166], suggest that predators like canids, herons, some spiders, some falcons, and several cetaceans usually hunt in a collaborative group, when it is less efficient to hunt alone (for instance, when the prey is large or can move as fast as the predators or when there are multiple prey). The nature of cooperation among individuals in the

© Springer International Publishing AG 2017
K.N. Kaipa and D. Ghose, *Glowworm Swarm Optimization,*
Studies in Computational Intelligence 698, DOI 10.1007/978-3-319-51595-3_7

group may range from simultaneous chases to hunts that are tightly coordinated [59]. Packer and Ruttan [166] show that the size and abundance of prey strongly influence the expected behavior of individuals in the hunting groups. They develop simple game-theoretic models that illustrate the conditions under which cooperative hunting occurs and how such cooperation can be detected. Gazda et al. [59] study cooperative hunting behavior of two groups of bottlenose dolphins and show that these cetaceans meet the criteria for the hypothesis that some predators adopt division-of-labor with role-specialization during group-hunting. These behaviors indicate that moving in formations may benefit predators during group hunting. Also, sensing limitations may require the members in the group to maintain tight spacing between each other [169]. These biologically-inspired notions can inspire synthetic rules that enable a collection of robotic pursuers to chase multiple mobile signal sources.

7.2 GSO Framework for Moving Sources

As mentioned earlier, unlike in the problem of source localization where the sources are stationary, the problem of cooperative pursuit involves mobile sources. Therefore, it is necessary to take into account the speed of movement of the glowworms (pursuers) and that of the moving source in analyzing the performance of GSO algorithm for mobile sources. It is assumed that the source radiates a signal whose intensity peaks at the source location and decreases monotonically with distance from the source. The moving radiation model assumes that the change in signal value at any point, due to the movement of source, is instantaneous. When a source is stationary, GSO ensures that the glowworms converge to the source location. When the source starts moving at relatively low speeds compared to that of the glowworms, it can be expected that the glowworms still converge to the source and move such that the centroid of the glowworms maintains the same position as the moving source. However, when the speed of the moving source is relatively high, there will be no convergence to the source point but rather a group of glowworms will move along with the source.

A scenario is considered where the glowworms pursue the source on a line (the positions of all the glowworms and the moving source are collinear), and present a theoretical result that provides a bound on the speed of the moving source below which the glowworms succeed in capturing the source.

Theorem 7.1 *Let V_p and V_e be the speeds of the glowworms and the moving source, respectively. If the positions of all the pursuers and the moving source are collinear, then a sufficient condition that enables the pursuers to capture the moving source is given by $V_e < 0.5V_p$.*

Proof Initially, it is shown that as the source moves, although the absolute values of luciferin-levels of the glowworms decrease as a function of the speed of the source, the relative order of luciferin-levels remain the same. For simplicity, consider a source

Fig. 7.1 Intensity of radiation signal at pursuers' locations as a function of time

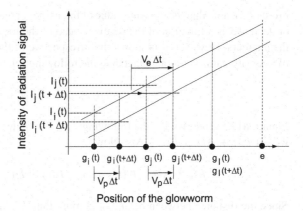

that radiates a signal I whose intensity linearly decreases with a slope m as we move away from the source. Consider three glowworms g_i, g_j, and g_l as shown in Fig. 7.1. It is assumed that the distance between adjacent neighbors is greater than the distance moved by each agent in one step ($= V_p \Delta t$).

Let the luciferin-levels of g_i, g_j, and g_l at time t, represented by $\ell_i(t)$, $\ell_j(t)$, and $\ell_l(t)$, satisfy the following relation:

$$\ell_i(t) < \ell_j(t) < \ell_l(t) \tag{7.1}$$

From (7.1), it is clear that g_i moves a distance $V_p \Delta t$ toward g_j, while g_j moves a distance $V_p \Delta t$ toward g_l. However, g_l remains stationary as it is the pursuer that is nearest to the source at time t.

Accordingly, the intensities at the pursuers' locations are updated as given below:

$$I_i(t + \Delta t) = I_i(t) + m(V_p - V_e)\Delta t$$
$$I_j(t + \Delta t) = I_j(t) + m(V_p - V_e)\Delta t$$
$$I_l(t + \Delta t) = I_l(t) - mV_e\Delta t \tag{7.2}$$

From (7.2), it is clear that

$$V_p > V_e \Rightarrow I_k(t + \Delta t) > I_k(t), \text{ for } k = i, j$$
$$V_p = V_e \Rightarrow I_k(t + \Delta t) = I_k(t), \text{ for } k = i, j$$
$$V_p < V_e \Rightarrow I_k(t + \Delta t) < I_k(t), \text{ for } k = i, j \tag{7.3}$$

However,

$$I_l(t + \Delta t) < I_l(t), \text{ for all } V_p \tag{7.4}$$

From (7.3), note that the absolute value of intensities at the pursuer-locations increase or decrease as a function of the relative speed of the moving source with respect to the pursuers. However, it is shown that irrespective of the speeds, the relative order of intensity values do not change and the following relation holds:

$$I_l(t + \Delta t) > I_j(t + \Delta t) > I_i(t + \Delta t). \tag{7.5}$$

Equation (7.5) is derived in the following way. Using the first two equations in (7.2), we get $I_j(t + \Delta t) > I_i(t + \Delta t)$. Now,

$$I_l(t + \Delta t) - I_j(t + \Delta t) = I_l(t) - I_j(t) - mV_p\Delta t \tag{7.6}$$

Since the distance between g_j and g_l is more than $V_p\Delta t$, we have $I_l(t) > I_j(t) + mV_p\Delta t$. Therefore, (7.6) becomes

$$I_l(t + \Delta t) > I_j(t + \Delta t)\Delta t \tag{7.7}$$

Using (2.1), we get

$$\begin{aligned}
\ell_i(t + \Delta t) &= (1 - \rho)\ell_i(t) + \gamma I_i(t + \Delta t) \\
\ell_j(t + \Delta t) &= (1 - \rho)\ell_j(t) + \gamma I_j(t + \Delta t) \\
\ell_l(t + \Delta t) &= (1 - \rho)\ell_l(t) + \gamma I_l(t + \Delta t)
\end{aligned} \tag{7.8}$$

Now, substituting from (7.8) and using (7.1) and (7.5),

$$\ell_j(t + \Delta t) - \ell_i(t + \Delta t) = (1 - \rho)(\ell_j(t) - \ell_i(t)) + \gamma(I_j(t + \Delta t) - (I_i(t + \Delta t)) \tag{7.9}$$
$$> 0$$

Similarly, it can be shown that

$$\ell_l(t + \Delta t) > \ell_j(t + \Delta t) \tag{7.10}$$

From (7.9) and (7.10) we have

$$\ell_l(t + \Delta t) > \ell_j(t + \Delta t) > \ell_i(t + \Delta t) \tag{7.11}$$

Therefore, from (7.11), although the absolute values of the luciferin levels of the pursuers change as a function of the relative speed of the moving source with respect to the pursuers, the relative order of luciferin levels is maintained irrespective of the speed of the moving source and hence, the pursuers move with a constant speed V_p until they lie outside a distance of $V_p\Delta t$ units from the pursuer (for instance, g_l in Fig. 7.1) that is nearest to the source.

Now, in order to determine a bound on the speed of the moving source below which the pursuers succeed in capturing the source, consider two pursuers g_j and

Fig. 7.2 Pursuer-source engagement scenario where the positions of the pursuers g_j and g_l and the moving source e are collinear. **a** $d_{jl} > V_p \Delta t$. **b** $d_{jl} < V_p \Delta t$

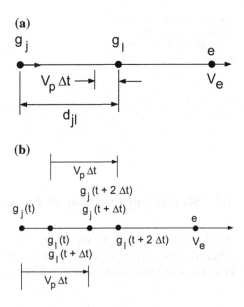

g_l such that their positions are collinear with respect to the source position. The pursuers-source engagement scenario is shown in Fig. 7.2a. Note that the distance between g_j and g_l, represented by d_{jl} in the figure, is more than $V_p \Delta t$. Therefore, g_j moves a distance $V_p \Delta t$ at every step and hence its effective speed is simply V_p.

However, once g_j moves to a point such that it is within a distance of $V_p \Delta t$ from g_l, the leap-frogging phenomenon of the GSO algorithm ensues and the effective speed of each glowworm is reduced to half of the original speed, which can be explained in the following way. In Fig. 7.2b, at time t, g_j leap-frogs over g_l and moves a distance $V_p \Delta t$. At time $t + \Delta t$, since g_j is closer to the moving source, g_j remains stationary and g_l leap-frogs over g_j and moves a distance $V_p \Delta t$. Therefore, each pursuer travels a net distance of $V_p \Delta t$ in every two time steps (equal to a duration of $2\Delta t$ units). Thus, the effective speed of each pursuer V_p^{eff} is given by

$$V_p^{eff} = 0.5 V_p \tag{7.12}$$

The speed of pursuer g_j as a function of its distance to g_l is shown in Fig. 7.3.

Now, using (7.12), the sufficient condition that enables the pursuers to capture the moving source is given by $V_e < 0.5 V_p$.   □

Fig. 7.3 Effective speed of
the pursuer g_j as a function
of its distance to the other
pursuer g_l

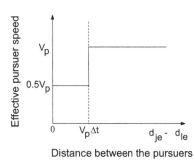

7.3 Simulation Experiment Results

A set of four simulation experiments are conducted in order to demonstrate the
efficacy of the algorithm for cooperative pursuit problems. The following function
(7.13) is used to model the signal radiation:

$$I(x, y) = \sum_{i=1}^{Q} a_i \exp(-b_i((x - x_i)^2 + (y - y_i)^2)) \tag{7.13}$$

where, Q represents the number of mobile sources and (x_i, y_i) represents the location
of each source. Note that the function $I(x, y)$ (7.13) is similar as $J_2(x, y)$ (3.1) with
the exception that each (x_i, y_i), $i = 1, \ldots, Q$, is a function of time. It is assumed
that the propagation delay is negligible and the local signal intensity is instantaneous.
Let $V_p^i = V_p$, $i = 1, \ldots, n$ (where n is the number of pursuers) be the speed of any
pursuer and let ψ be ratio of the speed of the moving source to the velocity of
the pursuer. Therefore, ψV_p gives the velocity of each mobile signal source. The
kinematic model of a moving source i is given below:

$$\dot{x}_i = \psi V_p \cos \theta_i \tag{7.14}$$
$$\dot{y}_i = \psi V_p \sin \theta_i \tag{7.15}$$
$$\dot{\theta}_i = \Omega_i \tag{7.16}$$

where, θ_i and Ω_i are the heading angle and angular velocity of the moving source i,
respectively.

For the purpose of experiments, it is assumed that the source moves in a circular
trajectory of radius r (Fig. 7.4).

Therefore we get,

$$\Omega_i = \frac{\psi V_p}{r} \tag{7.17}$$

Fig. 7.4 Mobile source
moving in a circular
trajectory with center at
(0, 0) and radius r

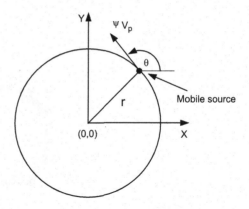

7.3.1 Simulation Experiment 1: Single Source Moving in a Circular Trajectory with Constant Angular Speed

A single mobile signal source that moves in a circular trajectory of radius 2 units and at a constant angular speed is considered. A set of 10 pursuers are randomly deployed in a workspace of size $(-5, 5) \times (-5, 5)$ units. The values of algorithmic constants used for the simulations are shown in Table 7.1. The emergence of pursuer movements and eventual chasing of the moving sources for two speed ratios ($\psi = 0.1, 0.2$) are shown in Figs. 7.5 and 7.6, respectively. The sources are captured in 324 and 149 iterations, respectively. The number of iterations required for source-capture for various values of ψ and for ten different experimental trials are shown in Table 7.2. Figure 7.7 shows the average number of iterations for source-capture as a function of ψ. Note from Table 7.2 that when $\psi \leq 0.4$, the source is captured in all the ten cases. However, when $\psi = 0.5$ and 0.6, the source is captured in three cases and one case, respectively and when $\psi \geq 0.7$, the source is never captured. From the above results it is clear that $0.4V_p$ acts as an upper bound on the speed of the moving source below which the pursuers succeed in chasing and capturing the source.

Table 7.1 Values of algorithmic constants used in the simulations

ρ	γ	s	β	n_t
0.4	0.6	0.3	0.08	2

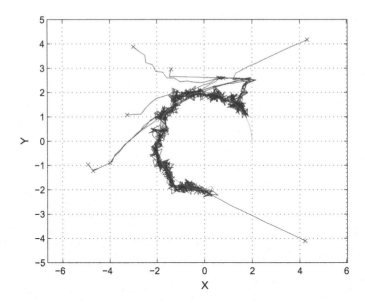

Fig. 7.5 Emergence of pursuer movements when the source is moving in a circular trajectory with center at $(0, 0)$ and radius $r = 2$ and $\psi = 0.1$

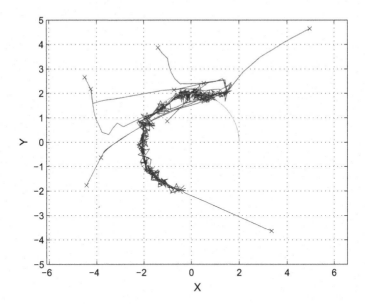

Fig. 7.6 Emergence of pursuer movements when the source is moving in a circular trajectory with center at $(0, 0)$ and radius $r = 2$ and $\psi = 0.2$

Table 7.2 No. of iterations for source-capture for various values of ψ

ψ | No.of trials	1	2	3	4	5	6	7	8	9	10
0.1	44	40	52	45	43	40	101	324	31	124
0.2	74	57	149	47	39	49	50	44	49	174
0.3	295	70	77	115	97	49	80	58	65	73
0.4	191	193	78	93	27	95	160	107	89	224
0.5	–	–	–	202	–	62	146	–	–	–
0.6	161	–	–	–	–	–	–	–	–	–
0.7	–	–	–	–	–	–	–	–	–	–

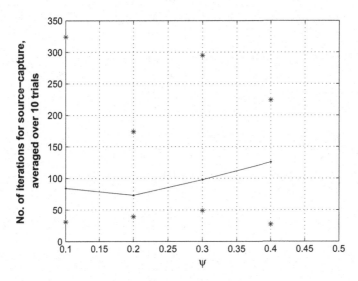

Fig. 7.7 Number of iterations for source-capture as a function of ψ, averaged over 10 trials. The maximum and minimum values of number of iterations for each value of ψ are marked as $*$'s

7.3.2 Simulation Experiment 2: Two Sources Moving in Two Overlapping Circular Trajectories

Two sources that move in circular trajectories of equal radii ($r = 2$) and centered at $(-1, 0)$ and $(1, 0)$, respectively, are considered. The sources move at a constant angular speed such that $\psi = 0.3$ ($\Rightarrow \dot{\theta} = \frac{0.3 \times 0.3}{2} = 0.015\,\text{rad/sec}$). The peak and slope values of the radiation functions for the two sources are chosen as $a_i = 3$, $i = 1, 2$ and $b_i = 0.1, i = 1, 2$. A set of 50 pursuers is randomly deployed in a workspace of $(-5, 5) \times (-5, 5)$ units (Fig. 7.8a). The sources are deployed at locations $(3, 0)$ and $(-3, 0)$, respectively. Snapshots of the pursuer and source positions after every 20 iterations are shown in Fig. 7.8b–i. Note that at $t = 40$, the swarm

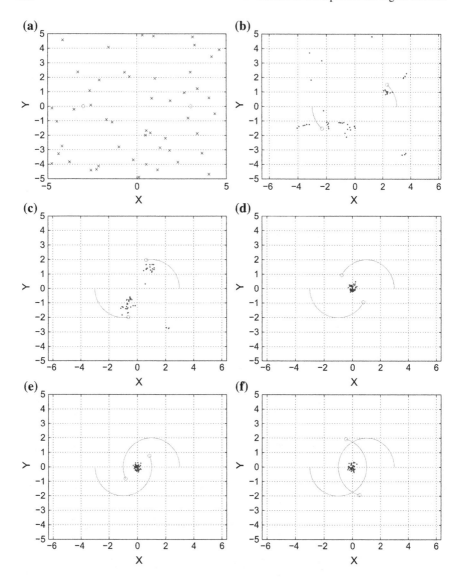

Fig. 7.8 Snapshots of positions of pursuers and sources at different time instants when the sources move in overlapping circles and for the case $b_i = 0.1$, $i = 1, 2$: **a** $t = 0$. **b** $t = 20$. **c** $t = 40$. **d** $t = 60$. **e** $t = 80$. **f** $t = 100$. **g** $t = 110$. **h** $t = 120$. **i** $t = 140$

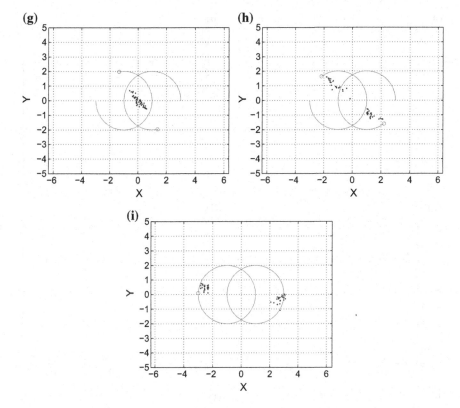

Fig. 7.8 (continued)

splits into two subgroups and each one of them pursues one of the two sources. However, between $t = 60$ and $t = 100$, the sources come close to each other leading to an increase in interference between their respective radiated signals. This causes the pursuers to move to the middle of the two sources until a time is reached when the sources start moving apart beyond a threshold distance. This situation can be observed at $t = 110$, when the swarm again splits into two subgroups as the sources move away from each other leading to a decrease in the interference levels. However, note that the interference level between radiation signals is a function of factors such as the intensity decay rate (b_i) and the distance between the two sources. For instance, It is observed in simulations that b_i can be increased to a value such that once the splitting of the swarm occurs, and each subgroup pursues one of the sources, the subgroups do not rejoin for all future time. The simulation result for the case $b_1 = b_2 = 0.6$ is shown in Fig. 7.9.

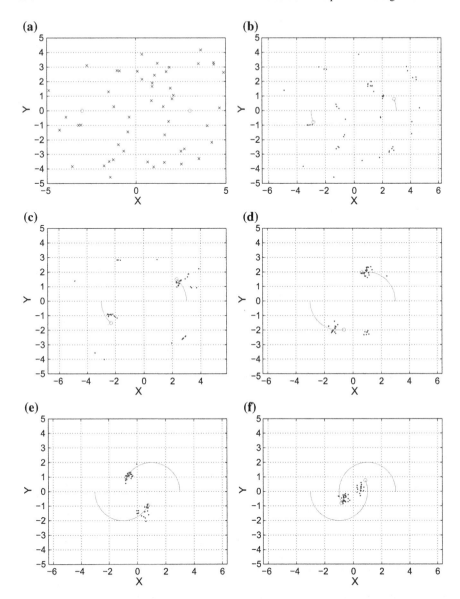

Fig. 7.9 Snapshots of positions of pursuers and sources at different time instants when the sources move in overlapping circles and for the case $b_i = 0.6, i = 1, 2$: **a** $t = 0$. **b** $t = 10$. **c** $t = 20$. **d** $t = 40$. **e** $t = 60$. **f** $t = 80$. **g** $t = 100$. **h** $t = 120$. **i** $t = 140$

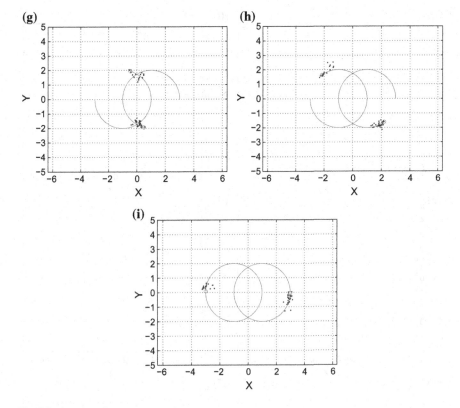

Fig. 7.9 (continued)

7.3.3 Simulation Experiment 3: Two Sources Moving in Concentric Circular Trajectories

Two sources that move in concentric circular trajectories centered at $(0, 0)$ and of radii 2 and 4, respectively, are considered. The sources move at constant and different angular speeds such that $\psi = 0.3$ ($\Rightarrow \dot{\theta}_1 = \frac{0.3 \times 0.3}{2} = 0.015$ rad/sec and $\dot{\theta}_2 = \frac{0.3 \times 0.3}{4} = 0.0075$ rad/sec). The peak and slope values of the radiation functions for the two sources are chosen as $a_i = 3, i = 1, 2$ and $b_i = 0.6, i = 1, 2$. A set of 50 pursuers is randomly deployed in a workspace of $(-5, 5) \times (-5, 5)$ units (Fig. 7.10a). The sources are deployed at locations $(2, 0)$ and $(-4, 0)$, respectively. Snapshots of the pursuer and source positions after every 30 iterations are shown in Fig. 7.10b–i. Note that by the end of 60 iterations (Fig. 7.10d), two subgroups of pursuers are formed and from there onwards, each subgroup keeps chasing one of the two sources. At $t = 150$ (Fig. 7.10g), as the two sources pass close to each other, the pursuers show a tendency to form a single group. However, as the sources move away from

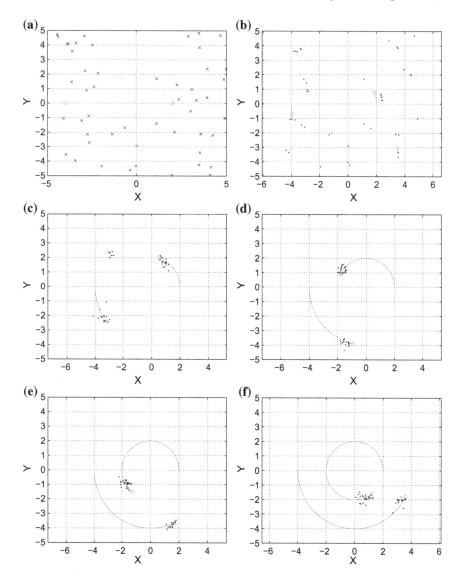

Fig. 7.10 Snapshots of positions of pursuers and sources at different time instants when the sources move in concentric circles and for the case $b_i = 0.6, i = 1, 2$: **a** $t = 0$. **b** $t = 10$. **c** $t = 30$. **d** $t = 60$. **e** $t = 90$. **f** $t = 120$. **g** $t = 150$ **h** $t = 180$. **i** $t = 210$

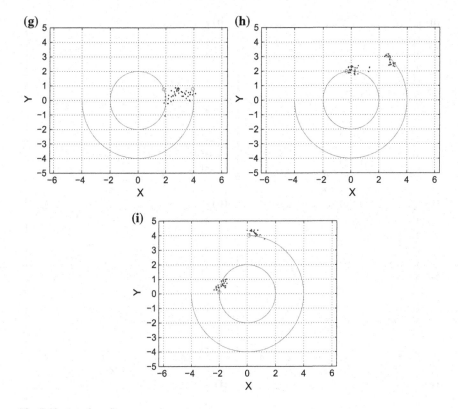

Fig. 7.10 (continued)

each other, the pursuers again split into two subgroups and resume the task of chasing their respective sources.

7.3.4 Simulation Experiment 4: Two Sources Moving Randomly About Nominal Concentric Circular Trajectories

Two sources that move randomly about nominal concentric circular trajectories centered at $(0, 0)$ and of radii 2 and 4, respectively, are considered. The random-movement of the sources is achieved by generating the radius of each source at each time t as given below:

$$r_1(t) = 2 + \vartheta$$
$$r_2(t) = 4 + \vartheta$$

where, ϑ is a uniformly distributed random variable in the interval $[0, 1]$.

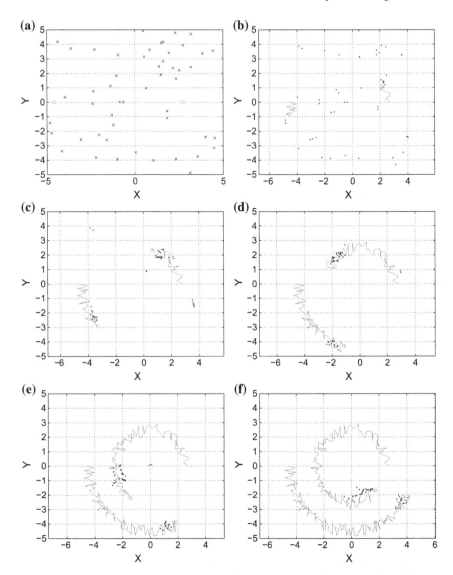

Fig. 7.11 Snapshots of positions of pursuers and sources at different time instants when the sources move randomly about nominal concentric circles and for the case $b_i = 0.6, i = 1, 2$: **a** $t = 0$. **b** $t = 10$. **c** $t = 30$. **d** $t = 60$. **e** $t = 90$. **f** $t = 120$. **g** $t = 150$ **h** $t = 180$. **i** $t = 210$

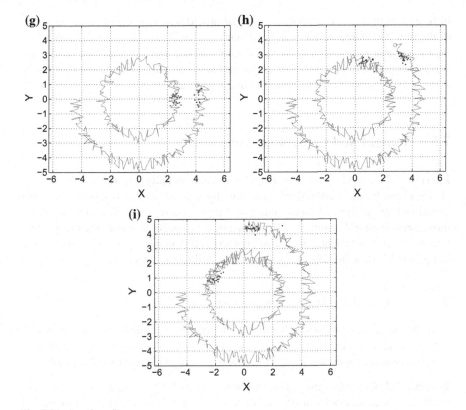

Fig. 7.11 (continued)

However, similar to the previous experiment, the sources move at constant and different angular speeds such that $\psi = 0.3$. The peak and slope values of the radiation functions for the two sources are chosen as $a_i = 3, i = 1, 2$ and $b_i = 0.6, i = 1, 2$. A set of 50 pursuers is randomly deployed in a workspace of $(-5, 5) \times (-5, 5)$ units (Fig. 7.11a). The sources are deployed at locations $(2, 0)$ and $(-4, 0)$, respectively. Snapshots of the pursuer and source positions after every 30 iterations are shown in Fig. 7.11b–i. Results show that the pursuers succeed even when some randomness is incorporated into the movements of the signal sources.

7.4 Summary

The problem of cooperative pursuit where a collection of mobile agents pursue multiple mobile signal sources was considered. The coordination scheme used by the pursuers was based on the GSO algorithm. The problem can be considered as a variation of the multiple source localization problem where the sources move instead of being stationary. A theoretical result that provides an upper bound on the speed of the moving source below which the pursuers succeed in chasing the source, in the case where the positions of the pursuers and the moving source are collinear, was presented. Simulation results demonstrated the efficacy of using the glowworm approach for the problem of chasing multiple sources.

7.5 Thought and Computer Exercises

Exercise 7.1 Consider two pursuer-glowworms, initially located at $(0, 0)$ and $(1, 0)$. Consider an evader-glowworm initially located at $(2, 0)$. Assume that the evader radiates a signal I whose intensity linearly decreases with a slope m as we move away from it. For time $t > 0$, assume that the evader moves along the positive x-axis. Let the movement step-size of pursuers and evader be s_p and s_e, respectively. Let $S_e = 0.12$. Run the GSO algorithm for the following cases:

1. $s_p = s_e$
2. $s_p > s_e$. Consider gradually increasing ratios of $\frac{s_p}{s_e} > 1$

In each case, determine whether or not the pursuers are able to capture the evader successfully. If not, explain why. If so, then determine the time to capture in each case. What is the minimum value of the ratio $\frac{s_p}{s_e}$ needed for successful capture.

Exercise 7.2 Consider three collinear pursuers at $(-1, 0)$, $(0, 0)$, and $(1, 0)$, respectively. Consider an evader at $(0, 1)$. For time $t > 0$, assume that the evader moves along the positive y-axis with a step-size $s_e = 0.12$. Will the pursuers succeed in capturing the evader? Provide a rationale for your answer.

Exercise 7.3 Consider three non-collinear pursuers at $(-1,0)$, $(0, 1)$, and $(1, 0)$, respectively. Consider an evader at $(0, 2)$. For time $t > 0$, assume that the evader moves along the positive y-axis with a step-size $s_e = 0.12$. Will the pursuers succeed in capturing the evader? Provide a rationale for your answer.

Exercise 7.4 Consider four pursuer-glowworms, initially located one at each corner of a square of side 2 units centered around $(0, 0)$. Let $s_e = 0.12$. Run the GSO algorithm to test capturability in the following cases:

1. A single evader-glowworm initially located at $(0, 2)$ that moves along the positive x-axis.
2. A single evader-glowworm initially located at $(0, 2)$ that moves along the positive y-axis.
3. A single evader-glowworm initially located at $(0, 0)$ that moves along the positive x-axis.

Choose an appropriate S_p for all these cases.

Exercise 7.5 Consider an initial random deployment of $n = 4$ glowworms in a square of side 2 units centered around $(0, 0)$. Run the GSO algorithm to test capturability in the three cases given in the previous exercise. Let $s_e = 0.12$. Repeat the capturability test for $n = 5, 6, \ldots, 10$. Report your observations on the effect of increasing n on capturability.

Exercise 7.6 Consider an initial random deployment of $n = 20$ glowworms in a square of side 2 units centered around $(0, 0)$. Consider two evaders that are initially located at $(0, 0)$. Consider one evader to move along the positive y-axis and the other evader to move along the negative y-axis. Determine suitable values of n and r_s that enable GSO to capture both the evaders.

Exercise 7.7 Consider an initial random deployment of $n = 20$ glowworms in a square of side 2 units centered around $(0, 0)$. Consider two evaders that are initially located at $(0, 0)$. Assume that the evaders perform a random walk. Determine suitable values of n and r_s that enable GSO to capture both the evaders.

Chapter 8
GSO Applications and Extensions

In this chapter, we present a survey on applications of GSO and its extensions. Work on GSO that appeared in the recent literature can be primarily classified into three categories. Researchers in the first category proposed modifications of GSO, which were mainly focused on either improving the convergence properties of original GSO or modifying GSO for global optimization problems. In the second category, researchers used basic GSO in different applications. In the third category, other researchers modified GSO and used them in some applications. Table 8.1 summarizes this classification.

8.1 Multiple Source Localization and Boundary Mapping

Localization of sources using mobile robot swarms has received some attention recently in the collective robotics community. Examples of such sources include sound, heat, light, leaks in pressurized systems [56], hazardous plumes/aerosols resulting from nuclear/chemical spills [227], fire-origins in forest fires [25], deep-sea hydrothermal vent plumes [88], hazardous chemical discharge in water bodies [53], oil spills [28], etc. This problem has also been recognized by the Department of Defense (DoD) as one of the applications that involves significant risks to humans [42].

Whereas most of the research in this area has dealt with single sources, relatively less research effort has been devoted to multiple source localization [104]. The problem is compounded when there are multiple sources. For instance, several forest fires at different locations give rise to a temperature profile that peaks at the locations of the fire. Multiple nuclear leaks and electromagnetic radiations originating at different locations can give rise to similar phenomenon. In all the above situations, there is an imperative need to simultaneously identify and neutralize all the sources before the emissions cause harm to the environment and people in the vicinity. In addition

© Springer International Publishing AG 2017
K.N. Kaipa and D. Ghose, *Glowworm Swarm Optimization*,
Studies in Computational Intelligence 698, DOI 10.1007/978-3-319-51595-3_8

Table 8.1 Classification of various works on GSO that appeared in the recent literature

Category I: Modifications of GSO
GSO-V1, GSO-V2, GSO-group, GSOV1C [163, 164]
Variable step-size and self-exploration [233]
Multi-population GSO [74]
Definite updating search domains [123]
Bioluminescent swarm optimization [37]
Improved GSO for high dimensional global optimization [234]
Niching GSO with mating behavior [85]
GSO with local search operator [235]
BFGS-GSO for global optimization [165]
GSO-Lèvy flights for global optimization [243]
GSO with elitism for global optimization [50]
Quantum GSO based on chaotic sequences [171]
AFSA and DE incorporated into GSO for global optimization [244]
Hierarchical multi-subgroups based GSO [76]
Parallelized GSO using MapReduce (programming model developed by Google) [3]
Improved GSO based on parallel hybrid mutation [203]
Other GSO variants with improved rules [75, 77, 219]
Category II: Applications of Basic GSO
Annual crop planning [27]
Nonlinear fixed charge transportation in a single stage supply chain network [135]
Cooperative swarm robotic exploration [33]
Optimization of polarimetric multiple-input multiple-output (MIMO) radar systems [90]
Design of web services [100]
Hydropower load allocation [214]
Category III: Applications of Modified GSO
Clustering [4, 85, 93, 242]
Wireless sensor networks [26, 121, 182, 211]
Multiple source localization and contaminant boundary mapping [140–142, 147, 148, 205, 206, 232]
Knapsack [65]
Numerical integration [223]
Solving fixed point equations [120]
Solving systems of nonlinear equations [120, 224]
Engineering design optimization [51, 188]
Environmental economic dispatch [89]
Parameter optimization of floatation processes [213]
Power generation optimization [144]
Unit commitment in power systems [116]
Whole-set order problem in scheduling [226]
Parameter optimizatin of Kaplan turbine [235]

to this, mapping of the contaminant boundary facilitates a rapid planning effort to move people and valuable property out of the affected region [147].

However, this problem raises the question of how to automatically partition the robots into subgroups in order to ensure that each source is captured by one of the subgroups. Other challenges, that are also common to the single source localization problem, include reaching a source in minimal time, obstacle avoidance with other robots during the course of the search, and how to contain an emission source after its detection.

In Chap. 1, an overview of prior research in the area and how GSO can be applied to address the multiple source localization problem were described. Here, we describe how other researchers modified original GSO to address the problems of source localization and boundary mapping.

Thomas and Ghose [205, 206] proposed a swarm algorithm that intelligently combines chemotactic, anemotactic, and spiralling behaviors in order to locate multiple odor sources. The chemotactic behavior was achieved by using GSO. Agents switch between the three behaviors based on the information available from the environment for optimal performance. The proposed algorithm was achieved by incorporating the following modifications into the GSO framework.

In turbulent flows, the peak concentration value within a patch and the frequency of encountering a patch increases as the glowworm gets closer to the source [10]. However, the instantaneous value might mislead the movement decision of the glowworms. Therefore, the authors defined the luciferin of each glowworm to be the maximum odor concentration encountered in the last N_{mem} seconds in its trajectory. This change was seen to improve algorithmic performance significantly. Accordingly, the luciferin update equation was modified as below:

$$\ell_i(t) \leftarrow \max\{C(x_i(t - N_{mem} + 1)), \ldots, C(x_i(t))\} \qquad (8.1)$$

where, $C(x_i(t))$ is the instantaneous odor concentration at glowworm i's location at time t (Fig. 8.1).

A glowworm without a neighbor, but with a nonzero luciferin value switches to the anemotactic behavior: it takes a step in the upwind direction, as given by (8.2), only when the measured concentration is above a threshold. This condition prevents a glowworm from leaving the plume and proceeding upwind away from the source. In case the concentration measured at its current position is below the threshold value, the glowworm stays at its current position.

$$x_i(t) \leftarrow x_i(t) - sw \qquad (8.2)$$

where, w is the wind direction and s is the step size.

A glowworm without a neighbor and with a zero luciferin value switches to a spiralling behavior until it either finds a neighbor or measures non-zero luciferin. The authors conducted a number of experiments to validate the algorithm's ability to simultaneously capture multiple odor sources. The proposed approach was later

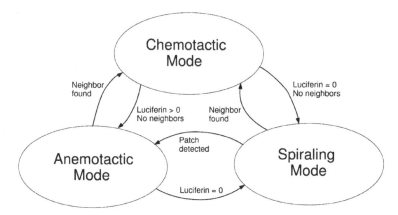

Fig. 8.1 Glowworm mode transition diagram

tested on data obtained from a dye mixing experiment. It was also seen capable of locating odor source under varying wind conditions.

A detailed survey on robot algorithms for localization of multiple emission sources was presented by McGill and Taylor [143]. The survey recognized that GSO directly addresses the issue of automatically partitioning a swarm into subgroups as required by the multiple source localization problem. This is achieved by the adaptive local-decision domain that facilitates the formation of subgroups in order to locate multiple sources simultaneously. In a comparative analysis, the authors reported that GSO can capture the highest number of sources (source no. 100, swarm size: 1000), when compared to other state-of-the-art algorithms like Biasing Expansion Swarm Approach (BESA) [36] (2, 20), Bayesian occupancy [186] (3, 20), Biased Random Walk (BRW) [41] (2, 100), PSO [138] (5, 10), and Attractant/repellant [61] (5, 11).

The authors conducted computer simulations based source localization experiments using GSO and compared its performance with BRW and a new GSO/BRW hybrid algorithm [141]. They considered a gradient field consisting of ten Gaussian sources on a continuous 1000×1000 unit space. Dead space was created by setting the field strength to zero if it was below a threshold value of 0.5. Three initial robot distributions were devised: *uniform* (agents are deployed at the node locations of a 2D grid spanning the search area), *drop* (all agents are deployed at a single location in the search area), and *line* (agents are deployed at one of the edges of the search area). BRW was used as a benchmark. A new variant GSO/BRW was proposed in which a glowworm without a neighbor performs a single BRW step; the glowworm resumes to act according to original GSO once it acquires at least one neighbor. It was shown that original GSO achieved maximum average number of sources and best convergence in the case of uniform distribution. GSO/BRW achieved better performance than GSO, and best performance was achieved by BRW, in the case of drop and line distributions. Similar computer simulations based comparative analysis was reported in [140]. The DIFFUSE algorithm was developed in [142] that addressed the problems of deadlocks. It achieved better performance in drop and line deployments. However, it was reported that the original GSO still achieved the best convergence performance in the case of uniform deployments (GSO: 100 vs GSO/BRW: 170 vs BRW: 2200 vs DIFFUSE: 2800 iterations for 95% convergence).

Later, Yuli et al. [232] showed that the limitations of GSO as described in [142] can be easily overcome by adding simple spiralling strategies (similar to [206]) and an adaptive step-size mechanism into the original GSO. In particular, they considered the same benchmark odor source problem used in [142] and showed that the modified GSO was able to capture all the ten sources for both drop and line distributions, yet converged much faster than the DIFFUSE algorithm. From these studies, we can conclude that suitable modifications to original GSO can greatly enhance its applicability to multiple emission source localization problems.

Methods till date have addressed the problems of either (multiple) source localization or boundary mapping [71] separately. However, in a recent contribution, Menon and Ghose [147] proposed a novel algorithm that enables a robotic swarm to achieve the following dual goals simultaneously: (1) localization of multiple sources of contaminants spread in a region and (2) mapping of the boundary of the affected region. The algorithm uses the basic GSO and modifies it considerably to make it suitable for both these tasks. Two types of agents, called the source localization agents (or S-agents) and boundary mapping agents (or B-agents) are used for this purpose. Whereas the S-agents behave according to the basic GSO, thereby solving the source localization problem, new behavior patterns are designed for the B-agents based on their terminal performance as well as interactions between them that help these agents to spread along the boundary to map its full length. Simulations were carried out on a static function profile with three contaminant sources with the assumption that the function values represent the time averaged contaminant intensities at that point. Results show that the algorithm effectively performs simultaneous source localization and boundary mapping.

8.2 Wireless Sensor Networks

Wireless sensor networks (WSNs) are large collections of sensor nodes with capabilities of perception, computation, communication, and locomotion. They are usually deployed in outdoor fields to carry out tasks like climate monitoring, vehicle tracking, habitat monitoring, earthquake observation, and surveillance. The performance of a WSN is mainly influenced by its coverage of the service area. This problem deals with finding an efficient deployment of the sensor nodes so that every location in the region of interest is sampled by a minimum of one node.

Deployment schemes can be broadly classified into two types: deterministic [215] and random [174]. Deterministic schemes are not feasible when sensors operate in dynamic environments and when *a priori* information is not available. Random deployment is more flexible as the sensor nodes can be disseminated conveniently without any prior information about the monitoring environment. Although feasi-

ble and easy to implement, random deployment cannot always guarantee complete
coverage and connectivity of the service area.

Liao et al. [121] proposed a sensor deployment scheme based on GSO that maximizes the coverage of the sensors with limited movement after an initial random
deployment. The decentralized nature of the GSO based approach leads to scalable
WSNs. They presented simulation results to show that their approach outperforms the
virtual force algorithm (VFA) [245] in terms of coverage rate and sensor movement.

They modeled the sensor deployment problem in the framework of GSO as
follows. Each sensor node is considered as a glowworm emitting luciferin whose
intensity is a function of its distance from its neighbors. Each glowworm has a
sensing range r_s and a communication radius r_c. In original GSO, a glowworm
is attracted toward a neighbor of brighter luminescence. On the contrary, in the
proposed approach, a glowworm is attracted toward its neighbors having dimmer
luminescence and decides to move toward one of them. These local movement rules
enable the glowworms to gradually distribute themselves within the sensing field so
that coverage is maximized. The luciferin of each glowworm i at time t is computed
as below:

$$\ell_i(t) = \ell_i(t-1) + \sum_{j=1}^{|N_i(t)|} \frac{\ell_j(t)}{d_{ij}^2(t)} \tag{8.3}$$

During the movement phase, each glowworm selects to move toward a neighbor
with the following probability:

$$p_{ij}(t) = \frac{\ell_i(t) - \ell_j(t)}{\sum_{k \in N_i(t)} \ell_i(t) - \ell_k(t)} \tag{8.4}$$

$$j \in N_i(t) \neq \phi$$

$$N_i(t) = \{j : d_{ij}(t) < r_c \text{ and } \ell_j(t) < \ell_i(t)\} \tag{8.5}$$

where $N_i(t)$ is the set of neighbors of glowworm i at time t.

The optimal distance between neighboring sensors for maximum coverage is $\sqrt{3}r_s$
[228]. Therefore, the distance moved by the glowworm toward its neighbor is chosen
as $\frac{\sqrt{3}r_s - d_{ij}(t)}{2}$. Therefore, the movement update for glowworm i is given by:

$$x_i(t+1) = x_i(t) + \left(\frac{\sqrt{3}r_s - d_{ij}(t)}{2}\right)\left(\frac{x_j(t) - x_i(t)}{||x_j(t) - x_i(t)||}\right) \tag{8.6}$$

The performance of the proposed GSO-based sensor deployment scheme was
evaluated by simulations. The authors considered a variable number of sensor nodes
($n = 50$, 100, and 200) with two different initial deployments: center and random.
A 2D obstacle-free environment was considered for deployment of sensors. The first
result showed that the coverage rate increased with an increase in number of nodes

and the GSO-based scheme achieved a higher coverage (96% for n = 200) compared to VFA (40% for n = 200) in all the cases. The second result showed that the coverage rate is higher in case of random deployment than center deployment irrespective of the number of sensors in the network. The third result showed that the GSO-based scheme achieved a lower moving distance compared to VFA.

Ray and De [182] applied a modified GSO to mobile sensor networks with the dual objectives of coverage and energy conservation. The authors solved the problem of maximizing sensor coverage, while minimizing the energy required to move the sensors to optimum locations. The solution is given in terms of deciding the minimum number of sensors to move and the total distance travelled. A modification of the basic GSO, called multi-parameter reverse GSO, was proposed. Wang et al. [211] used GSO for wireless sensor networks to optimize energy consumption and prolong the life time of the network. This work combined GSO with clustering technique and sink mobility strategies. Chen et al. [26] proposed a hybrid algorithm composed of both GSO and PSO to solve the problem of wireless sensor network coverage under intrusion attacks.

8.3   Clustering

The problem of clustering deals with partitioning a set of objects into clusters so that the objects in the same cluster are more similar to each other than to those in other clusters according to some similarity measure. The K-means clustering method is one of the simplest unsupervised learning algorithms for solving the clustering problem [128]. Clustering has important applications in exploratory data analysis, pattern recognition, machine learning, and other engineering fields. Two clustering techniques that were developed based on GSO are presented here.

8.3.1   CGSO

Kao et al. [93] modeled the data clustering problem as a continuous optimization problem and solved it by using GSO. Experimental results showed that GSO based clustering (CGSO) is very competitive compared to other meta-heuristic approaches. They formulated the clustering problem as a minimization problem:

$$\text{Min} \sum_{i=1}^{K} \sum_{j=1}^{n_i} ||x_j - m_i||, \sum_{i=1}^{K} n_i = n \tag{8.7}$$

$$||x_j - m_i|| = \sqrt{\sum_{p=1}^{P} (x_{jp} - m_{ip})^2} \tag{8.8}$$

where x_j is the jth data point, P is the number of data attributes, n_i is the number of data points in the ith cluster, and m_i is the ith cluster-center.

Each data point x_j corresponds to the position of a glowworm j. Now, the fitness at each glowworm's position is computed as below:

$$J(x_j) = \min ||x_j - m_i||, i = 1, 2, \ldots, K \qquad (8.9)$$

The luciferin value of each glowworm was computed such that a better objective function value translated to a higher luciferin value. The CGSO algorithm was tested on three data clustering problems and the results were compared with the K-means and the ACO based clustering method (SACO) proposed by Shelokar [193]. The three data sets—Iris, Thyroid, and Wine—were selected from the UCI repository of machine learning databases. Experimental results showed that CGSO achieved better clustering performance when compared to K-means and SACO on all the three data sets. Although K-Means required the shortest computational time in all three cases, it provided worse objective values. CGSO spent shorter time, yet found better solution quality, than SACO for all three data sets.

8.3.2 GSO for Self-organized Clustering

In another recent study, Huang and Zhou [85] showed that GSO can be used to perform self-organized data clustering by suitably defining the fitness function. Whereas the number of clusters K was explicitly specified in CGSO, its value emerged as a result of the self-organizing ability of GSOCA. Similar to CGSO, the position of each glowworm i represents a cluster data object $X(x_1, x_2, \ldots, x_m)$ in the m−dimensional search space. The fitness at each glowworm's position is computed as below:

$$J(X) = -ln\left(\frac{1}{n}\right) + ln(\delta(X)) \qquad (8.10)$$

where $\delta(X) = \frac{|N(X,r)|}{n}$ is the local space relative density and $N(X, r)$ is the data points contained in the local space within a distance r of X. Now, the luciferin of each glowworm is computed by using (8.10) in the original luciferin update rule defined in Chap. 2. The rest of the algorithm remains the same as described in Chap. 2. It was shown that repeated application of GSO enabled the data objects to eventually self-organize into clusters that respected the pattern similarity constraints.

According to (8.10), note that a data object near the centroid of a cluster has a higher luciferin value than that of data objects at a cluster boundary. This results in the automatic splitting of the data objects and association of data objects with the centroid of their respective clusters. The main advantage of GSOCA is the self-organized clustering without a specification of the number of clusters and its ability to deal with non-convex data sets.

Next, the authors combined GSOCA with K-means clustering in order to improve the clustering performance: GSOCA is implemented for a short period in order to achieve automatic clustering. This results in spherical, or close to spherical shaped clusters, which is used as an initial seed by a K-means algorithm for further refinement of clusters.

The authors considered three different data sets to test their clustering approach:

1. A 2-D problem with four unique classes consisting a total of 600 patterns drawn from four independent bivariate normal distributions
2. A 3-D problem with five classes consisting a total of 250 patterns
3. The iris data set (a popular benchmark in pattern recognition and machine learning) consisting of one linearly separable and two linearly non-separable classes

In all these test cases, the authors showed that GSOCA+K-means outperformed basic K-means, PSO, and two other clustering algorithms (*New 1* and *New 2*) proposed in [92].

8.4 Knapsack Problem

Knapsack is a well known problem in combinatorial optimization. Given a set of items, each with a weight and a value, we need to find the number of each item to fill into a group so that the total value is maximized and the total weight is not greater than a given limit. In a 0–1 knapsack problem, the number of each item can be either 0 or 1. Gong et al. [65] proposed a variation of GSO to solve a multidimensional 0–1 knapsack problem (MKP). The MKP is an extension of the standard knapsack problem in which the weight is a vector valued variable (representing a set of resources, for example). The problem is formulated as follows:

$$\text{Max} \sum_{j=1}^{m} x_j c_j \tag{8.11}$$

$$\text{s.t.} \sum_{j=1}^{m} a_{ij} x_j \leq b_i, i = 1, 2, \cdots, g \tag{8.12}$$

$$x_j \in \{0, 1\}, j = 1, 2, \cdots, m \tag{8.13}$$

where m is the number of items, $x_j \in \{0, 1\}$ represents whether item j is included in the group or not, c_j is the profit made by item j, g is the number of resources, a_{ij} is the item j's consumption of resource i, and b_i is the maximum total limit on resource i.

The GSO formulation to solve the MKP is briefly described as follows. Let the location of glowworm i be $x_i = \{x_{i1}, x_{i2}, \ldots, x_{im}\}$, $x_{ij} \in \{0, 1\}$. The fitness at glowworm i's location is chosen as

$$J(x_i) = \sum_{j=1}^{m} x_{ij} c_j \tag{8.14}$$

$$\text{s.t.} \sum_{j=1}^{m} a_{ij} x_j \le b_i, i = 1, 2, \ldots, g \tag{8.15}$$

At the beginning of every GSO cycle, a greedy local search is performed for each glowworm with an attempt to increase the fitness, while maintaining feasibility of the solution. The remaining steps of the cycle are updated according to the equations of original GSO. The authors reported results from tests on ten instances of multidimensional knapsack problems in order to validate the efficacy of their approach.

8.5 Numerical Integration

Numerical integration is a standard problem in engineering/scientific calculations and deals with computing an approximate solution to a definite integral. If the integrand is a smooth well-behaved function and the limits are bounded, there are many methods like Newton method, Gauss method, Romberg method, and Simpson's method that approximate the integral with arbitrary precision. However, the problem is difficult to tackle when the antiderivative for the integrand is not easy to obtain.

Yang et al. [223] applied GSO for computing numerical integrals. They showed that the GSO based numerical integration algorithm has an ability to not only compute usual definite integrals, but also compute singular integrals and definite integrals in which the antiderivative of the integrand is not an elementary function.

Let the definite integral be defined as $\int_a^b f(x)dx$, where $f(x)$ is the integrand. Each glowworm i is considered as a string of d nodes that are randomly deployed between the left and right endpoints of the integral interval. This results in a total of $d+2$ nodes and the division of the interval into $d+1$ sections. Let $(x_i^{(1)}, x_i^{(2)}, \ldots, x_i^{(d)})$ be the set of positions of the d nodes of glowworm i.

Next, the set of integrand values at the $d + 2$ nodes and at the midpoint of each section are computed, giving rise to a total of $2d + 3$ values:

$$F_i = \left\{ f(a), f\left(\frac{a + x_i^{(1)}}{2} \right), f(x_i^{(1)}), f\left(\frac{x_i^{(1)} + x_i^{(2)}}{2} \right), f(x_i^{(2)}), \right.$$
$$\left. \ldots, f(x_i^{(d)}), f\left(\frac{x_i^{(d)} + b}{2} \right), f(b) \right\}$$

Next, the maximum and minimum of each section are found as below:

$$W_i^{(j)} = \max\left\{ f(x_i^{(j)}), f\left(\frac{x_i^{(j)} + x_i^{(j+1)}}{2}\right), f(x_i^{(j+1)}) \right\} \qquad (8.16)$$

$$w_i^{(j)} = \min\left\{ f(x_i^{(j)}), f\left(\frac{x_i^{(j)} + x_i^{(j+1)}}{2}\right), f(x_i^{(j+1)}) \right\} \qquad (8.17)$$

where, $j = 0, 1, 2, \ldots, d + 1$; $x_i^{(0)} = a$; $x_i^{(d+2)} = b$

Now, the fitness of each glowworm i is computed as below:

$$J(i) = \frac{1}{2}\sum_{j=1}^{d+1} \delta_j |W_j - w_j| \qquad (8.18)$$

where δ_j is the length of section j. Note that as the value of $J(i)$ reaches closer to zero, the fitness of glowworm becomes better.

The value of $J(i)$ is used in the luciferin update equation of original GSO to determine the luciferin values of all glowworms at each iteration. The rest of the algorithm is implemented according to original GSO. After the termination condition is reached, the best glowworm i^* is chosen to compute the numerical integral by using the expression:

$$\sum_{j=1}^{d+1} f\left(\frac{x_{i^*}^{(j-1)} + x_{i^*}^{(j)}}{2}\right)\delta_j \qquad (8.19)$$

The authors tested the validity of using GSO to compute numerical integrals on a set of nine functions with varying degrees of complexity—usual integrands, singular integrands, and integrands whose antiderivates are not elementary functions. The results presented by the authors, summarized in Table 8.2, reveal that GSO compares well with the performance of other methods like trapezoid method, Simpson's method, neural networks, ES, and Rectangle method.

8.6 Fixed Point Equations

Fixed point theory can be applied to problems like algebraic equations, differential equations, integral equations, and stability analysis that are frequently encountered in mathematics and control engineering. A fixed point equation can be defined as follows: suppose X is a nonempty set and $T : X \to X$. If $\exists x^* \in X$ such that $T(x^*) = x^*$, then x^* is called a fixed point of T [91]. Currently, cauchy iteration method (CIM) [231] is often used to solve fixed point equations. However, the drawbacks of CIM include the need to verify the conditions of Banach theorem [158], lack of parallelism, only one solution at a time, and heavy influence of initial point on the efficiency and accuracy of the solution.

Table 8.2 Numerical integration results obtained by using GSO and other algorithms (Results reported by Yang et al. [223])

Definite integral	Accurate value	GSO	Trapezoid method	Simpson's method	Neural networks	ES	Rectangle method
$\int_0^2 \sqrt{1+x^2}\,dx$	2.958	2.822	3.326	2.964	–	–	–
$\int_0^2 x^2 dx$	2.667	2.337	4.000	2.667	–	–	–
$\int_0^2 x^4 dx$	6.400	5.080	16.000	6.667	–	–	–
$\int_0^2 \frac{1}{1+x}dx$	1.099	0.991	1.333	1.111	–	–	–
$\int_0^2 \sin x\,dx$	1.416	1.415	0.909	1.425	–	–	–
$\int_0^2 e^x dx$	6.389	5.405	8.389	6.421	–	–	–
$\int_0^{48} \sqrt{1+\cos^2 x}\,dx$	58.470	58.460	–	58.471	58.520	58.471	–
$e^{-x},\ 0 \le x < 1$ $e^{-x/2},\ 1 \le x < 2$ $e^{-x/3},\ 2 \le x < 3$	1.5460	1.5402	–	–	1.5467	1.5459	–
$\int_0^1 e^{-x^2}dx$	0.7468	0.7478	0.7462	0.7468	–	0.7468	0.7778

Table 8.3 Error in the solution to the fixed point equations obtained by three methods (Results reported by Qu et al. [120])

	Iterations	CIM	GSO	Hybrid GSO
$x = \ln(x) + 2$ in $[2, \infty)$	20	3.200542053 605204e-010	6.2729729 27541254e-004	0
$\begin{cases} 2x_1 - x_2 = 4 \\ -2x_1 + 3x_2 = 6 \end{cases}$	100	2.603995296 937122e-010	0.00424020 8500842	2.512147933 894040e-015

Qu et al. [120] transformed the fixed point equation into an optimization problem and applied GSO to solve it. They also proposed a hybrid GSO, which embeds CIM in the original GSO for local search in order to improve solution accuracy. The GSO based formulation to solve fixed point equations is discussed below.

The solution x^* to the fixed point equation $T(x^*) = x^*$ is the same as the solution to the following optimization problem:

$$\text{Min } ||x - T(x)||_2 \tag{8.20}$$

where $||.||_2$ represents the 2-norm.

Based on (8.20), the fitness of each glowworm i can be defined as follows:

$$J(x_i) = \frac{1}{1 + ||x - T(x)||_2} \tag{8.21}$$

Now, the original GSO is implemented with a minor modification in the movement update rule: At iteration t, if $N_i(t) = \phi$, then $x_i(t + 1) = T(x_i(t))$. This enables a neighborless glowworm to implement a local search by using the Cauchy iteration procedure. The results reported by the authors on two test problems, summarized in Table 8.3, show that GSO outperforms CIM and the best performance is achieved by the hybrid version of the algorithm.

8.7 System of Nonlinear Equations

Many fundamental problems in engineering and science can be formulated as systems of nonlinear equations. Finding solutions to these problems can be very challenging and computationally expensive. Moreover, selection of initial values is difficult for most traditional algorithms. Yang et al. [224] developed a modified GSO algorithm to solve a system of nonlinear equations represented as below:

$$\begin{cases} f_1(x_1, x_2, \ldots, x_n) = 0 \\ f_1(x_1, x_2, \ldots, x_n) = 0 \\ \quad \vdots \\ f_m(x_1, x_2, \ldots, x_n) = 0 \end{cases} \tag{8.22}$$

Solving the system of nonlinear equations in (8.22) can be formulated as follows:

$$\text{Let } J(X) = \sum_{i=1}^{m} f_i^2(X) \text{where, } X = (x_1, x_2, \ldots, x_n) \tag{8.23}$$

Now, a solution to (8.22) is given by X^* that satisfies $J(X) = 0$. The value of X^* can be determined by solving the following optimization problem.

$$X^* = \arg\{\min J(X)\} \tag{8.24}$$

Now GSO is applied to the above problem by using $J(X)$ as the fitness function. An additional local search based on Hooke-Jeeves pattern search method [82] is introduced into the algorithm. After the movement update of each glowworm, its current optimal location and the corresponding optimal value are set as the initial values of the local search method. After the local search for all glowworms is completed, the algorithm returns to the next cycle of GSO. It is observed that the hybrid GSO achieves an accuracy of 10^{-15} in finding the approximate roots of equations for a set of six test problems. For three of these problems, the hybrid GSO achieves better accuracy than that of other algorithms from literature. For two of these problems, the hybrid GSO achieves better accuracy than that of PSO. The interested reader is referred to [224] for more details on the analysis of the results.

In a similar work, Qu et al. [120] embedded simplex search [162] into each GSO cycle and applied the modified GSO to the problem of solving a system of nonlinear equations. Whereas the local search was performed for all glowworms in [224], it was performed only for the glowworms without neighbors, that have higher chances of being located at a local optimum. For the same set of six test problems used in [224], simplex-GSO achieved better accuracy than that of Hooke–Jeeves-GSO and other algorithms from literature. The interested reader is referred to [120] for more details on the analysis of the results. Zhou et al. [240] presented a method to solve nonlinear equations using a modification of GSO called the Leader-GSO, in which at the end of one generation, the best glowworm is designated a leader and all the other glowworms are made to move toward this leader.

8.8 Engineering Design Optimization

8.8.1 Active Vibration Control of Smart Structures

Smart structures refer to self-regulating structures that use embedded actuators/sensors and feedback control mechanisms in order to sense, and adjust, their behavior as conditions change [125]. Large cantilever beams prone to externally induced vibrations (e.g., aircraft wings, rotor blades, space manipulators, etc.) are suitable candidates that can be designed by using smart structure technologies. Piezoelectric materials are usually used as actuators and sensors in these applications. Active vibration control of smart structures involves an optimal placement of these devices and optimal tuning of feedback gains; this translates to a maximization of the energy dissipated by the underlying feedback control system.

Dutta et al. [51] applied a variation of GSO to find solutions to the above optimization problem. Initially, they mathematically proved that the system is uncontrollable if the actuators and sensors are placed at the nodal points of the mode shapes. Based on this result, they set the actuator/sensor locations and the feedback gain matrix as design variables and formulated the problem of finding their optimal values as solving a constrained nonlinear optimization problem. Next, they converted this problem into an unconstrained optimization problem by using penalty functions. Next, they applied a variation of GSO to compute solutions to this continuous optimization problem in a six dimensional search space (2 location and 4 gain variables). As the global optimum was sought, the neighborhood range was enhanced to cover the entire search space.

First, the authors considered a cantilever beam with one and two collocated actuator(s)/sensor(s) and presented the numerical results obtained by using GSO. Next, they considered an extension of this cantilever beam problem with five collocated actuators/sensors and presented similar results obtained by using GSO. They also investigated the effect of increasing the number of design variables on the optimization process. Experimental comparisons demonstrated that the performance of GSO was better than multistart quasi-Newton [114] and float encoded GA [229] and was on par with artificial bee colony (ABC) [95]. These results show that GSO is a good choice for the optimization of smart structures.

Similarly Wang et al. [212] used discrete GSO to solve the problem of sensor and actuator placement, posed as a combinatorial optimization problem, for vibration control.

8.8.2 Composite Laminates Buckling Load Maximization

Composite laminates are materials formed by joining layers of fibrous composites in different configurations so as to provide properties—high stiffness, high strength-to-weight ratios, long fatigue life, resistance to electro-chemical erosion strength, and desired coefficient of thermal expansion–that are critical to materials used in industries like aerospace, marine, and automobile manufacturing [11, 63, 136]. The goal of achieving the desired material properties can be formulated as an optimization problem, in which optimal values of design variables are sought that optimize a suitably defined objective.

Scimemi and Rizzo [188] solved the problem of buckling load maximization in composite laminates by using GSO. The primary reason that motivated the authors to apply GSO to this problem is its ability to provide alternative solutions that are theoretically sub-optimal, but possibly optimal when the real world constraints are taken into account. The authors considered a composite laminate formed by 64 plies, in which each ply is a graphite epoxy lamina with constant elastic properties and thickness. Therefore, the only design variables that influence the buckling load factor are the fiber orientations of each lamina. Three different sets of fiber directions and constraints (symmetric and balanced, symmetric only) were considered that gave rise to 16, 32, and 32 design variables, respectively. Experimental results showed that GSO was able to capture the distinct global optima (9, 32, and 32) in all the three test cases.

8.9   Variants of GSO

8.9.1   GSO-v1, GSO-v2, and GSO-Group

In original GSO, a glowworm does not change its position until it finds at least one neighbor in its current neighborhood. Oramus [163] proposed three variations of GSO–GSO v1, GSO v2, and GSO group–that aim to eliminate situations in which a glowworm does not move.

GSO-v1: A glowworm with an empty neighborhood moves to a trial location χ and the fitness at the new location is evaluated. The move is accepted only if the solution doesn't worsen; otherwise, the glowworm's position remains unchanged. This is equivalent to one typical cycle in a hill climbing algorithm. Accordingly, the movement update rule of the original GSO is modified as below.

$$\forall i : N_i(t) = \emptyset \text{ do} \tag{8.25}$$

$$\chi = x_i(t) + s(0.5 - R) \tag{8.26}$$

$$x_i(t + 1) = \begin{cases} \chi, & \text{if } J(\chi) \geq J(x_i(t)) \\ x_i(t), & \text{otherwise} \end{cases} \tag{8.27}$$

where R is a random number having a uniform distribution in the range $[0, 1)$.

GSO-v2: In addition to the hill climbing feature for every glowworm without a neighbor, the sensor range r_s is extended by a factor of 5% when number of glowworms inside a sphere of radius r_s is less than the desired number of neighbors n_d.

This modification improves the chances of neighborless glowworms in finding neighbors and weakens the effect of random movements introduced by the first modification.

GSO-group: Glowworms always move and collaborate in predefined groups. In each group there is one *master* glowworm that is placed in the middle of the group, and several *slave* glowworms that are uniformly distributed on a hyper-sphere centered at master's position. The master determines a common movement direction for all the members of the group. The distance between master and slaves is initially equal to 10% of the sensor range. The master-slave distance for a given group is controlled by the master and varies in the range from 33% of the step size to 10% of the sensor range. In each iteration, the master checks the luciferin value of its own slaves and in most cases uses the original GSO to designate the movement direction of a common group step. Members from other groups are queried only in the case when all group members have the same value of the luciferin.

Tests on the Peaks function revealed that when $n = 25$, the performance of all three GSO variants were better than that of original GSO. When $n = 100$, only GSO-group was significantly better in comparison to original GSO. The performance of GSO-v2 was worse than that of original GSO. On the Himmelblau's function, the GSO variants performed better than GSO when $n = 25$. The performance of all four algorithms were equally good when $n = 100$.

Oramus [164] developed another variant of GSO called GSOV1C by adding a feature of "conservation of movement directions" to GSO-v1. In particular, when a glowworm has multiple directions to move, it selects a direction that deviates less from its movement direction in the previous step with a higher probability. Further performance improvement was reported due to this modification.

8.9.2 Variable Step Size (V-GSO) and Self-exploration (E-GSO)

The step size s is kept fixed in original GSO. Zhang et al. [233] proposed V-GSO, in which a variable step size strategy is used to improve the convergence speed of GSO. Two variable step-size strategies were studied:

Linear decreasing strategy. The step size s is decreased linearly with time according the following equation:

$$s(t) = (s_0 - s_{min})\frac{(T_{max} - t)}{T_{max}} + s_{min} \tag{8.28}$$

where, s_0 and s_{min} are the initial and termination values of s, respectively ($s_0 > s_{min}$); T_{max} is the maximum number of iterations.

Nonlinear decreasing strategy. The step size s is decreased nonlinearly with time. Two curves—a parabola with the convex side facing upward (8.29) and a parabola with the convex side facing downward (8.30)—are considered:

$$s(t) = (s_{min} - s_0) \left(\frac{t}{T_{max}} \right)^2 + s_0 \qquad (8.29)$$

$$s(t) = (s_0 - s_{min}) \left(\frac{t}{T_{max}} \right)^2 + (s_{min} - s_0) \left(\frac{2t}{T_{max}} \right) + s_0 \qquad (8.30)$$

The authors performed tests on a set of three benchmark multimodal functions and showed that all the three variable step-size strategies increased the speed of convergence by at least 28% in comparison with fixed step-size strategy. Next, the authors added a self-exploration behavior to V-GSO, giving rise to E-GSO: When the fitness of a glowworm's current location is lower than a given threshold and it has no neighbor whose fitness is above a given threshold, the glowworm performs either a random spiral search or a Z-shaped search. Further improvement in convergence speed was observed when E-GSO was tested on the same set of benchmark functions.

8.9.3   Multi-population GSO (MPGSO)

He and Zhu [74] presented multi-population GSO (MPGSO), a variation of GSO based on small scale and multiple populations. The principal modification in MPGSO is achieved by partitioning the search space into k roughly equal regions and deploying a population of $\frac{n}{k}$ glowworms in each region. However, the details of how to choose the value of k are not given. The solutions found by all the populations are merged to find the total set of solutions over the entire search space. In examples of two multimodal test functions, the Peaks function and Random peaks function, the authors showed that MPGSO can capture higher number of peaks in reduced time. However, tests on more benchmark functions are needed to validate the performance of MPGSO.

8.9.4   Definite Updating Search Domains (GSO-D)

In original GSO, a glowworm doesn't move until it finds at least one neighbor in its current neighborhood. Liu et al. [123] proposed a minor modification to this rule: During the movement phase, each glowworm without a neighbor moves closer to the glowworm with the best fitness. The modified movement update rule is given as below.

$$x_i(t+1) = \begin{cases} x_{best}(t+1) + (rand - 0.5), & \text{if } N_i(t) = \emptyset \\ x_i(t) + s \left(\frac{x_j(t) - x_i(t)}{\|x_j(t) - x_i(t)\|} \right), & \text{otherwise} \end{cases} \qquad (8.31)$$

where x_{best} is the position of the glowworm with the highest fitness.

On a set of eight benchmark test functions borrowed from [39], the authors showed that the above modification of definite movement updating improved the global search accuracy of GSO.

8.9.5 Bioluminescent Swarm Optimization Algorithm (BSO)

Oliveira et al. [37] proposed bioluminescent swarm optimization (BSO), a new variant of GSO that is geared for global optimization problems. BSO was obtained by adding the following new features to the original GSO: stochastic adaptive step sizing, global optimum attraction, leader movement, and mass extinction. The first two features are implemented by modifying the original movement update rule as below:

$$x_i(t+1) = x_i(t) + rand\ s \left(\frac{x_j(t) - x_i(t)}{\|x_j(t) - x_i(t)\|} \right) + c_g\ rand\ s \left(\frac{g(t) - x_i(t)}{\|g(t) - x_i(t)\|} \right)$$

where, $s = s_0 \frac{1}{1+c_s \ell_i(t)}$, s_0 is the maximum step size, $g(t)$ is the global best position, c_g is the global best attraction constant, and c_s is a slowing constant. Leader movement was achieved by performing a weak local search (using local unimodal sampling [170]) every iteration and a strong local search (using single-dimension perturbation search) every few iterations on $g(t)$. Mass extinction is usually used in nature-inspired algorithms as a mechanism to prevent premature stagnation. This was implemented in BSO by reinitializing the positions of all particles, except $g(t)$, if there was no improvement in $g(t)$ after a fixed number of iterations. The authors compared the performance of BSO with PSO. On a set of four benchmark test functions, BSO achieved better accuracy than PSO. The convergence was slower than that of PSO, thereby avoiding early stagnation and leading to better solutions.

Other variants of GSO in the recent literature include improved GSO for high-dimensional global optimization [234], niching GSO with mating behavior [85], and GSO with a local search operator [235].

8.10 Thought and Computer Exercises

Note: In this chapter, we studied several examples of GSO's application to problems in the fields of machine learning, optimization, and robotics. These problems included clustering, knapsack, numerical integration, solving fixed point equations, solving systems of nonlinear equations, engineering design optimization, wireless sensor networks, multiple source localization, and boundary mapping. Accordingly, the quesitons in this chapter will be based on how GSO can be customized to these problems.

Exercise 8.1 Define clustering. Give some examples where clustering can be applied. Provide a mathematical formulation of the clustering problem. K-means algorithm, a conventional technique to solve the clustering problem, is given as follows. Given a set of points $(x_1, x_2, \ldots, x_n)$, where x_i is an m-dimensional vector, find a partition $S = \{S_1, S_2, \ldots, S_k\}$ that minimizes the following objective function:

$$\sum_{i=1}^{k} \sum_{x \in S_i} ||x - \mu_i||^2 \tag{8.32}$$

where μ_i is the mean of points in S_i

Write a Matlab code to implement the K-means and run it on three data sets—Iris, Thyroid, and Wine—selected from the UCI repository of machine learing database.[1] Comment on your results.

Exercise 8.2 Describe how GSO is suitable to solving clustering problems. Three algorithms—clustering GSO, self-organized GSO, and self-organized GSO with K-means—were described in Sect. 8.3 to solve the problem of clustering. Provide the algorithmic steps in terms of pseudocode for each case. Write the corresponding Matlab codes and run the algorithms on the same data sets used in Exercise 8.1. Use these results to provide a comparitive analysis of the three algorithms and the standard K-means algorithm. Use solution quality and computational time as the performance metrics during your analysis.

Exercise 8.3 Define the knapsack problem and give examples where it can be used. Provide the mathematical representation for the general multidimensional knapsack problem (MKP). From Sect. 8.4, note that the search space used for solving the MKP is a discrete binary space. However, the original GSO was formulated for continuous spaces. What modifications in the various steps of the GSO algorithm are needed for application to discrete spaces.

Exercise 8.4 The GSO variant described in Sect. 8.4 defines the glowworm location, specifies a fitness function, and combines greedy search and original steps of GSO to solve the MKP. Use this information to enumerate the algorithmic steps of the GSO variant in terms of pseudocode. Translate this into Matlab code and run the algorithm on the seven test problems in "mknap1.txt" found in this link.[2]

Exercise 8.5 What is numerical integration? What are difficulties faced when approximating integrals? Formulate a variant of GSO that can perform numerical integration (Hint: Use the method described in Sect. 8.5 as a reference). Use this method to find approximate values of the following definite integrals:

---

[1] https://archive.ics.uci.edu/ml/datasets.html.

[2] http://people.brunel.ac.uk/~mastjjb/jeb/orlib/mknapinfo.html.

$$\text{(a)} \int_0^2 \sqrt{1+x^2}dx$$

$$\text{(b)} \int_0^2 x^2 dx$$

Exercise 8.6 The GSO variant described in Sect. 8.5 can also compute singular integrals and definite integrals in which the antiderivative of the integrand is not an elementary function. Apply the GSO variant to the following singular integral and verify if this is true.

$$f(x) = \begin{cases} \exp^{-x}, 0 \le x < 1 \\ \exp^{-x/2}, 1 \le x < 2 \\ \exp^{-x/3}, 2 \le x < 3 \end{cases}$$

Now, consider the following integral where the integrand is a non-elementary function. Apply the GSO variant and verify if it can approximate the integral.

$$I = \int_0^1 \exp^{-x^2} dx$$

Exercise 8.7 Define the problem of fixed point equation and provide its mathematical representation. How can it be framed as an optimization problem? Work out the steps of GSO that can be used to solve this problem based on the methods described in Sect. 8.6. Run a Matlab code to solve the fixed point of the following nonlinear equation and report the error in the resulting solution:

$$x = \ln(x) \in [2, +\inf)$$

Exercise 8.8 Write a Matlab code for the hybrid GSO described in Sect. 8 and run it on the above test functions. Compare the results of hybrid GSO with that of basic GSO.

Exercise 8.9 Suppose (X, d) is a complete metric space and T satisfies the conditions of Banach fixed point theorem. That is, $T : X \longrightarrow X$ is a contraction mapping. Therefore, T has a unique fixed point in X. Now, the Cauchy-iteration method (CIM) to solve the fixed point equation is given as follows. Let $x^{(0)}$ be a arbitrarily chosen initial value.

$$x^{(1)} = T(x^{(0)}) \tag{8.33}$$
$$x^{(2)} = T(x^{(1)}) \tag{8.34}$$
$$\cdots \tag{8.35}$$
$$x^{(i)} = x^{(i-1)} \tag{8.36}$$

When i is large enough, $x^{(i)}$ can be seen as an approximate solution of the fixed point equation. Run the CIM on the test function given in Exercise 8.7 and compare the results with that of GSO and hybrid GSO.

Exercise 8.10 Formulate the problem of solving a system of nonlinear equations using the GSO framework. Consider the following system of nonlinear equations:

$$= \begin{cases} f_1(x) = x_1^2 + x_2 + 1 = 0 \\ f_2(x) = x_1 - \cos(0.5\pi x_2) = 0 \end{cases}$$

where $x \in [-2, 2]$. The exact roots of this system lie at $x^* = (-1/\sqrt{(2)}, 1.5), x^* = (0, 1)$. Apply the GSO formulation to solve the above set of equations. Report the accuracy of the resulting solution.

Propose methods to augment GSO's capability by using Hooke-Jeeves pattern search and simplex search. Check whether or not the solution accuracy improves by using these two hybrid versions of GSO.

Appendix A
GSO Code in MATLAB

```
%----------------GSO.m----------------------------------------------------
% Glowworm swarm optimization (GSO)
% Developed by K.N. Kaipa and D. Ghose in 2005
% This is the main front-end code
%-------------------------------------------------------------------------

clc; clear all; close all;

tic

global n m A_init A Ell gamma ro step1 r_d r_s ...
       beta r_min n_t Ave_d bound

m = 2; % No. of dimensions

% Parameter initialization
%-----------------------------------------------
n     = 100;               % No. of agents
r_s   = 3;                 % Sensor range
r_d   = r_s*ones(n,1);     % Local decision range
r_min = 0;                 % Threshold decisin range
gamma = 0.6;               % Luciferin enhancement constant
ro    = 0.4;               % Luciferin decay constant
step1 = 0.03;              % Distance moved by each glowworm when a
                           % decision is taken
beta  = 0.08;              % decision range gain
n_t   = 5;                 % Desired no. of neighbors

% Initialization of variables
%-----------------------------------------------------
bound = 3;                 % Parameter specifying the workspace range
DeployAgents;              % Deploy the glowworms randomly
Ell = 5*ones(n,1);         % Initialization of Luciferin levels
j = 1;                     % Iteration index
iter = 250;                % No. of iterations
Ave_d = zeros(iter,1);     % Average distance

% Main loop
%-----------------------------------------------
while (j <= iter)
    UpdateLuciferin;       % Update the luciferin levels at glowworms'
                           % current positions
    Act;                   % Select a direction and move
```

© Springer International Publishing AG 2017
K.N. Kaipa and D. Ghose, *Glowworm Swarm Optimization,*
Studies in Computational Intelligence 698, DOI 10.1007/978-3-319-51595-3

```
    for  k = 1 : n   % store the state histories
         agent_x(k,j,:) = A(k,1);
         agent_y(k,j,:) = A(k,2);
    end
    j = j + 1;
    j                       % Display iteration number
end

toc                        % Display the total computation time

% Plots
%-------------------------------------------------
figure(1);                 % Plot of trajectories of glowworms from their
                           % initial locations to final locations
plot(A_init(:,1),A_init(:,2),'x');
xlabel('X'); ylabel('Y');
hold on;
DefineAxis;
for  k = 1 : n
     plot(agent_x(k,:,:),agent_y(k,:,:));
end
DefineAxis;
grid on;
hold on;
plot([-0.0093;1.2857;-0.46],   [1.5814;-0.0048;-0.6292],'ok');

figure(2);                 % Plot of final locations of glowworms
plot(A(:,1),A(:,2),'.');
DefineAxis;
grid on;
hold on;
plot([-0.0093;1.2857;-0.46], [1.5814;-0.0048;-0.6292],'ok');

%Save each of the following 9 functions as a separate "FunctionName.m" file
%in the same folder as "GSO.m"

% Function 1: DeployAgents.m-----------------------------------------------

function DeployAgents

global n m A_init A bound

 B = -bound*ones(n,m);
 A_init = B + 2*bound*rand(n,m);

 A = A_init;

% Function 2: UpdateLuciferin.m--------------------------------------------

function UpdateLuciferin

global n A J Ell gamma ro

for i = 1 : n
     x = A(i,1); y = A(i,2);

     % The Matlab 'Peaks' function is used here. Please replace it with
     % the multimodal function for which peaks are sought

     J(i,:) = 3*(1-x)^2*exp(-(x^2) - (y+1)^2) ...
              - 10*(x/5 - x^3 - y^5)*exp(-x^2-y^2) ...
              - 1/3*exp(-(x+1)^2 - y^2);
```

```matlab
        Ell(i,:) = (1-ro)*Ell(i,:) + gamma*J(i,:);
end

% Function 3: Act.m-----------------------------------------------------

function Act

global n r_s r_d N N_a beta n_t

N(:,:) = zeros(n,n);
N_a(:,:)= zeros(n,1);

for i = 1 : n
    FindNeighbors(i);
    FindProbabilities(i);
    Leader(i) = SelectAgent(i);
end
for i = 1 : n
    Move(i,Leader(i));
    r_d(i) = max(0, min(r_s,r_d(i) + beta*(n_t-N_a(i))));
end

% Function 4: FindNeighbors.m-------------------------------------------------

function FindNeighbors(i)

global n m A N r_d N_a Ell

n_sum = 0;

for j = 1 : n
    if (j~=i)
        square_sum = 0;
        for k = 1 : m
            square_sum = square_sum + (A(i,k)-A(j,k))^2;
        end
        d = sqrt(square_sum);
        if (d <= r_d(i)) & (Ell(i) < Ell(j))
            N(i,j) = 1;
            n_sum = n_sum + 1;
        end
    end
    N_a(i) = n_sum;
end

% Function 5: FindProbabilities.m---------------------------------------

function FindProbabilities(i)

global n N Ell pb

Ell_sum = 0;
for j = 1 : n
    Ell_sum = Ell_sum + N(i,j)*(Ell(j) - Ell(i));
end
    if (Ell_sum == 0)
    pb(i,:) = zeros(1,n);
else
    for j = 1 : n
        pb(i,j) = (N(i,j)*(Ell(j)-Ell(i)))/Ell_sum;
    end
end
```

```
% Function 6: SelectAgent.m-----------------------------------------------

function j = SelectAgent(i)

global n pb

bound_lower = 0;
bound_upper = 0;
toss = rand;
j = 0;
for k = 1 : n
    bound_lower = bound_upper;
    bound_upper = bound_upper + pb(i,k);
    if (toss > bound_lower) & (toss < bound_upper)
        j = k;
        break;
    end
end

% Function 7: Move.m-----------------------------------

function Move(i,j)

global A m step1 Ell bound

if (j~=0) & (Ell(i) < Ell(j))
    temp(i,:) = A(i,:) + step1*Path(i,j);
    flag = 0;
    for k = 1 : m
        if (temp(i,k) < -bound) | (temp(i,k) > bound)
            flag = 1;
            break;
        end
    end
    if (flag == 0)
        A(i,:) = temp(i,:);
    end
end

% Function 8: Path.m------------------------------------

function Del = Path(i,j)

global A m

square_sum = 0;
for k = 1 : m
    square_sum = square_sum + (A(i,k)-A(j,k))^2;
end
hyp = sqrt(square_sum);

for k = 1 : m
    Del(:,k) = (A(j,k) - A(i,k))/hyp;
end

% Function 9: DefineAxis.m------------------------------------

function DefineAxis

global bound

axis([-bound bound -bound bound]);
grid on;
```

Appendix B
GSO Code in C++

```cpp
//----------------GSO.cpp----------------------------------------------
// Glowworm swarm optimization (GSO)
// Developed by K.N. Kaipa and D. Ghose in 2005
// This is the main front-end code
----------------------------------------------------------------------

#include <iostream>
#include<fstream>
#include <algorithm>
#include <math.h>

using namespace std;

/* Parameters

Constants
----------

n - Number of glowworms
r - Sensor range
rho - Luciferin decay constant
gama - Luciferin enhancement constant
beta - neighborhood ehancement constant
s - step size (agent speed)
nd - Desired number of neighbors
d - dimension of the search space


Variables
----------

Lc - Luciferin value
Rd - Neighborhood range
P - Probability matrix
Ld - Leader set
N - Neighborhood matrix
Na - Actual number of neighbors
X - Glowworm positions
W - Workspace size

*/

#define n 1000
```

© Springer International Publishing AG 2017

229

K.N. Kaipa and D. Ghose, *Glowworm Swarm Optimization,*
Studies in Computational Intelligence 698, DOI 10.1007/978-3-319-51595-3

```
#define r 125.0
#define rho 0.4
#define gama 0.6
#define beta 0.08
#define s 1
#define d 2
#define nd 5
#define PI 3.14159
#define W 3
#define IterMax 500

static int Ld[n], N[n][n], Na[n], randSeed = 1;
//static float s = 0.03;
static double X[n][d], Lc[n], Rd[n], P[n][n], Sol[] = {-PI/2, PI/2};
//{-0.5268, 0.5268};  //{-10, 5.24};

static ofstream outFile ("peaks_sol.dat",ios::app);
//static ofstream deploy ("InitialPos.dat",ios::app);

static double Distance(int i, int j) {
double dis = 0;
for(int k = 0; k < d; k++)
dis = dis + pow((X[i][k]-X[j][k]),2);
return sqrt(dis);
}

static void DeployGlowworms(float lim) {
            for(int i = 0; i < n; i++) {
                for(int j = 0; j < d; j++) {
                    X[i][j] = -lim + 2*lim*rand()/(RAND_MAX+1.0);
//std::cout << X[i][j] << "\n" ;
 }
// deploy << X[i][0] << ' ' << X[i][1] << "\n";
  }
  //deploy.close();

}


static void UpdateLuciferin() {
for (int i = 0; i < n; i++) {
double x = X[i][0], y = X[i][1];
//double J = - pow(x,2) - pow(y,2);
double J =  3*pow((1-x),2)* exp(-pow(x,2) - pow((y+1),2)) - 10 * (x/5 - pow(x,3) -
pow(y,5)) * exp(-pow(x,2) - pow(y,2)) -  1/3 * exp(-pow((x+1),2) - pow(y,2));
//double J = 0;
//for(int j = 0; j < d; j++) {
//J = J + pow(sin(X[i][j]),2);
//J = J + 10 + pow(X[i][j],2) - cos(2*PI*X[i][j]);
//J = J + 418.9829 + X[i][j]*sin(sqrt(fabs(X[i][j])));
//}
//cout << t << J << "\n";
//getchar();
Lc[i] = (1-rho)*Lc[i] + gama*J;
}
}


static void FindNeighbors() {

for(int i = 0; i < n; i++) {
N[i][i] = 0; Na[i] = 0;
for(int j = 0; j < n; j++){
```

```
if (j!=i){
if ((Lc[i] < Lc[j]) && (Distance(i,j) < Rd[i])) N[i][j] = 1;
else N[i][j] = 0;
Na[i] = Na[i] + N[i][j];
}
}
}
}

static void FindProbabilities() {
 for(int i = 0; i < n; i++) {
  double sum = 0;
  for (int j = 0; j < n; j++) sum = sum + N[i][j]*(Lc[j] - Lc[i]);
  for(int j = 0; j < n; j++) {
   if (sum != 0) P[i][j] = N[i][j]*(Lc[j] - Lc[i])/sum;
    else P[i][j] = 0;
//printf("%f ", P[i][j]);
  }
  //printf("\n");
 }
}

static void SelectLeader() {
  for (int i = 0; i < n; i++) {
 double b_lower = 0;
 Ld[i] = i;
  double toss = rand()/(RAND_MAX + 1.0);
  for (int j = 0; j < n; j++) {
   if (N[i][j] == 1) {
    double b_upper = b_lower + P[i][j];
    if ((toss >= b_lower) && (toss < b_upper)) {
     Ld[i] = j;
     break;
    } else b_lower = b_upper;
   }
  }
 // printf("%d ", Ld[i]);

 }

}

static void Move() {
 for (int i = 0; i < n; i++) {
  if (Ld[i]!=i) {
   int flag = 0;
   double temp[d];
   double dis = Distance(i,Ld[i]);
   //if (Na[i] > 15) s = 0.0001; else s = 0.03;
   for (int j = 0; j < d; j++) {
    temp[j] = X[i][j] + s*(X[Ld[i]][j] - X[i][j])/dis;
if (fabs(temp[j]) > W) {
 flag = 1;
 break;
}
   }
   if (flag == 0) for (int j = 0; j < d; j++) X[i][j] = temp[j];
  }
 }
}

static void UpdateNeighborhood() {
for (int i = 0; i < n; i++)
```

```cpp
  Rd[i] = max(0.0, min(r, Rd[i] + beta*(nd - Na[i])));
}

static void Peaks() {
int cnt = 0, pk_cnt = 0, pks[int (pow(3,d))], pk_sum = 0;
for(int i = 0; i < pow(3,d); i++)
 pks[i] = 0;
for (int a = 0; a < 3; a++) {
for (int b = 0; b < 3; b++) {
for (int c = 0; c < 3; c++) {
for (int y = 0; y < 3; y++) {
for (int e = 0; e < 3; e++) {
for (int f = 0; f < 3; f++) {
for (int g = 0; g < 3; g++) {
for (int h = 0; h < 3; h++) {
for (int i = 0; i < 3; i++) {
for (int j = 0; j < 3; j++) {
int cntr = 0;
for (int k = 0; k < n; k++) {

if (sqrt(pow(Sol[a]-X[k][0],2) + pow(Sol[b]-X[k][1],2)) < 1
// + pow(Sol[c]-X[k][2],2) + pow(Sol[y]-X[k][3],2)
   + pow(Sol[e]-X[k][4],2) + pow(Sol[f]-X[k][5],2)
   + pow(Sol[g]-X[k][6],2) + pow(Sol[h]-X[k][7],2)
   + pow(Sol[i]-X[k][8],2) + pow(Sol[j]-X[k][9],2)) < 1)
{
cntr++;
pks[cnt]++;
}
}
cnt++;
}
}
}
}
}
}
}
}
}
for (int i = 0; i < pow(3,d); i++) {
  pk_sum = pk_sum + pks[i];
  if (pks[i] > 2) pk_cnt++;
  cout << pks[i] << "\n";
}
cout << "\n" << cnt << ' ' << pk_cnt << ' ' << pk_sum << "\n";
} */

static void Peaks() {
int cnt = 0, pk_cnt = 0, pks[int (pow(2,d))], pk_sum = 0;
for(int i = 0; i < pow(2,d); i++)
 pks[i] = 0;
for (int a = 0; a < 2; a++) {
for (int b = 0; b < 2; b++) {
for (int c = 0; c < 2; c++) {
for (int e = 0; e < 2; e++) {
for (int f = 0; f < 2; f++) {
for (int g = 0; g < 2; g++) {
for (int h = 0; h < 2; h++) {
for (int z = 0; z < 2; z++) {
for (int k = 0; k < n; k++)
   if (sqrt(pow(Sol[a]-X[k][0],2) + pow(Sol[b]-X[k][1],2)
```

```
           + pow(Sol[c]-X[k][2],2) + pow(Sol[e]-X[k][3],2)
           + pow(Sol[f]-X[k][4],2) + pow(Sol[g]-X[k][5],2)
           + pow(Sol[h]-X[k][6],2) + pow(Sol[z]-X[z][7],2)) < s)
    pks[cnt]++;
    cnt++;
    }
    }
    }
    }
    }
    }
    }
    }
    //cout << cnt << ' ' << pk_cnt;
    for (int i = 0; i < pow(2,d); i++) {
      pk_sum = pk_sum + pks[i];
      if (pks[i] > 2) pk_cnt++;
        //cout << ' ' << i << ' ' << pks[i] << "\n";
    }
    //cout << "\n" << cnt << ' ' << pk_cnt << ' ' << pk_sum << "\n";
    outFile << cnt << ' ' << pk_cnt << ' ' << pk_sum << "\n";


    }


int main (int argc, char * const argv[]) {


    for (int currParam = 0; currParam < argc; currParam++) {

    if (strcmp(argv[currParam], "-r") == 0)
        randSeed = atoi(argv[currParam+1]);

    }
    cout << randSeed << "\n";
    srand(randSeed);
    DeployGlowworms(W);
    for (int i = 0; i < n; i++) {
    Lc[i] = 5; Rd[i] = r;
    }
    for (int t = 0; t < IterMax; t++ ) {
//cout << t << "\n";
UpdateLuciferin();
FindNeighbors();
FindProbabilities();
SelectLeader();
Move();
UpdateNeighborhood();
//printf("%f %f \n", X[0][0], X[0][1]);
//getchar();
    }
// Peaks();
    for(int i = 0; i < n; i++) {
    //cout << s*exp(-0.1*Na[i]) << "\n";
    outFile << i << ' ' << X[i][0] << ' ' << X[i][1] << "\n";
    }
    //getchar();
    outFile.close();
    exit(0);
    return 0;
}
```

Appendix C
Useful Links

Link to Wikipedia page on Glowworm Swarm Optimization:
   http://en.wikipedia.org/wiki/Glowworm_swarm_optimization

Link to Wikipedia page on Swarm Intelligence:
   http://en.wikipedia.org/wiki/Swarm_intelligence

Link to Wikipedia page on Evolutionary Multimodal Optimization:
   http://en.wikipedia.org/wiki/Evolutionary_multi-modal_optimization

Link to Wikipedia page on Metaheuristics:
   http://en.wikipedia.org/wiki/Metaheuristic

© Springer International Publishing AG 2017   235
K.N. Kaipa and D. Ghose, *Glowworm Swarm Optimization,*
Studies in Computational Intelligence 698, DOI 10.1007/978-3-319-51595-3

References

1. D.H. Ackley. An empirical study of bit vector function optimization. *Genetic algorithms and simulated annealing*, 1:170–204, 1987.
2. N. Al-Madi, I. Aljarah, and S.A. Ludwig. Parallel glowworm swarm optimization clustering algorithm based on mapreduce. In *Proceedings of the IEEE International Symposium on Swarm Intelligence*, pages 1–8, Dec 2014.
3. I. Aljarah and S.A. Ludwig. A mapreduce based glowworm swarm optimization approach for multimodal functions. In *Proceedings of the IEEE Symposium on Swarm Intelligence*, pages 22–31, April 2013.
4. I. Aljarah and S.A. Ludwig. A new clustering approach based on glowworm swarm optimization. In *Proceedings of the IEEE Congress on Evolutionary Computation*, pages 2642–2649, June 2013.
5. I. Aljarah and S.A. Ludwig. A scalable mapreduce-enabled glowworm swarm optimization approach for high dimensional multimodal functions. *International Journal of Swarm Intelligence Research*, 7(1):32–54, 2016.
6. H. Ando, Y. Oasa, I. Suzuki, and M. Yamashita. Distributed memoryless point convergence algorithm for mobile robots with limited visibility. *IEEE Transactions on Robotics and Automation*, 15(5):818–828, 1999.
7. J. Atema. Chemical signals in the marine environment: dispersal, detection, and temporal signal analysis. *Proceedings of the National Academy of Sciences*, 92(1):62–66, 1995.
8. B.G. Babu and M. Kannan. Lightning bugs. *Resonance*, 7(9):49–55, 2002.
9. T. Back, D.B. Fogel, and Z. Michalewicz, editors. *Handbook of Evolutionary Computation*. IOP Publishing Ltd., Bristol, UK, UK, 1st edition, 1997.
10. E. Balkovsky and B.I. Shraiman. Olfactory search at high reynolds number. *Proceedings of the National Academy of Sciences*, 99(20):12589–12593, 2002.
11. E.J. Barbero. *Introduction to Composite Materials Design*. CRC Press, 2010.
12. P. Barthelemy, J. Bertolotti, and D.S. Wiersma. A lévy flight for light. *Nature*, 453(7194):495–498, 2008.
13. M. Beekman, A.L. Gilchrist, M. Duncan, and D.J.T. Sumpter. What makes a honeybee scout? *Behavioral Ecology and Sociobiology*, 61(7):985–995, 2007.
14. R.A. Bernstein. Foraging strategies of ants in response to variable food density. *Ecology*, 56(1):213–219, 1975.
15. G. Bilchev and I.C. Parmee. The ant colony metaphor for searching continuous design spaces. In *Proceedings of the AISB Workshop on Evolutionary Computing*, pages 25–39, 1995.
16. K.G. Blair. Luminous insects. *Nature*, 96(2406):411–415, 1915.

K.N. Kaipa and D. Ghose, *Glowworm Swarm Optimization*,
Studies in Computational Intelligence 698, DOI 10.1007/978-3-319-51595-3

17. C. Blum and M. Dorigo. The hyper-cube framework for ant colony optimization. *IEEE Transactions on Systems, Man, and Cybernetics, Part B (Cybernetics)*, 34(2):1161–1172, 2004.
18. E. Bonabeau, M. Dorigo, and G. Theraulaz. *Swarm Intelligence: From Natural to Artificial Systems*. Oxford University Press, 1999.
19. E. Bonabeau, M. Dorigo, and G. Theraulaz. Inspiration for optimization from social insect behaviour. *Nature*, 406(6791):39–42, 2000.
20. R. Brits, A.P Engelbrecht, and F. Van den Bergh. A niching particle swarm optimizer. In *Proceedings of the 4th Asia-Pacific conference on simulated evolution and learning*, pages 692–696, 2002.
21. J. Buck and E. Buck. Biology of synchronous flashing of fireflies. *Nature*, 211(5043):562–564, 1966.
22. J.B. Buck. Synchronous flashing of fireflies experimentally induced. *Science*, 81(2101):339–340, 1935.
23. B. Bullnheimer, R.F. Hartl, and C. Strauss. A new rank based version of the ant system - a computational study. *Central European Journal for Operations Research and Economics*, 7:25–38, 1997.
24. G. Di Caro and M. Dorigo. Antnet: distributed stigmergetic control for communications networks. *Journal of Artificial Intelligence Research*, 9(1):317–365, 1998.
25. D.W. Casbeer, R.W. Beard, T.W. McLain, S-M. Li, and R.K. Mehra. Forest fire monitoring with multiple small uavs. In *Proceedings of the American Control Conference*, pages 3530–3535, June 2005.
26. Z. Chen, X. Li, B. Yang, and Q. Zhang. A self-adaptive wireless sensor network coverage method for intrusion tolerance based on trust value. *Journal of Sensors*, 2015, 2015.
27. S. Chetty and A.O. Adewumi. Comparison study of swarm intelligence techniques for the annual crop planning problem. *IEEE Transactions on Evolutionary Computation*, 18(2):258–268, 2014.
28. J. Clark and R. Fierro. Cooperative hybrid control of robotic sensors for perimeter detection and tracking. In *Proceedings of the American Control Conference*, pages 3500–3505, June 2005.
29. M. Clerc. *Particle Swarm Optimization*. John Wiley and Sons, 2010.
30. R. De Cock. Larval and adult emission spectra of bioluminescence in three european firefly species. *Photochemistry and Photobiology*, 79(4):339–342, 2004.
31. M.S. Couceiro, R.P. Rocha, and N.M.F. Ferreira. Ensuring ad hoc connectivity in distributed search with robotic darwinian particle swarms. In *Proceedings of the IEEE International Symposium on Safety, Security, and Rescue Robotics*, pages 284–289, Nov 2011.
32. M.S. Couceiro, R.P. Rocha, and N.M.F. Ferreira. A novel multi-robot exploration approach based on particle swarm optimization algorithms. In *Proceedings of the IEEE International Symposium on Safety, Security, and Rescue Robotics*, pages 327–332, Nov 2011.
33. M.S. Couceiro, P.A. Vargas, R.P. Rocha, and N.M.F. Ferreira. Benchmark of swarm robotics distributed techniques in a search task. *Robotics and Autonomous Systems*, 62(2):200–213, 2014.
34. I.D. Couzin and N.R. Franks. Self-organized lane formation and optimized traffic flow in army ants. *Proceedings of the Royal Society of London B: Biological Sciences*, 270(1511):139–146, 2003.
35. H. Cui, J. Feng, J. Guo, and T. Wang. A novel single multiplicative neuron model trained by an improved glowworm swarm optimization algorithm for time series prediction. *Knowledge-Based Systems*, 88:195–209, 2015.
36. X. Cui, C.T. Hardin, R.K. Ragade, and A.S. Elmaghraby. A swarm approach for emission sources localization. In *Proceedings of the IEEE International Conference on Tools with Artificial Intelligence*, pages 424–430, Nov 2004.
37. D.R. de Oliveira, R.S. Parpinelli, and H.S. Lopes. Bioluminescent swarm optimization algorithm. *Evolutionary Algorithms*, pages 69–84, 2011.
38. J. Dean and S. Ghemawat. Mapreduce: Simplified data processing on large clusters. *Communications of ACM*, 51(1):107–113, 2008.

39. K. Deep and J.C. Bansal. Mean particle swarm optimisation for function optimisation. *International Journal of Computational Intelligence Studies*, 1(1):72–92, 2009.
40. J.L. Deneubourg, S. Aron, S. Goss, and J.M. Pasteels. The self-organizing exploratory pattern of the argentine ant. *Journal of Insect Behavior*, 3(2):159–168, 1990.
41. A. Dhariwal, G.S. Sukhatme, and A.A.G. Requicha. Bacterium-inspired robots for environmental monitoring. In *Proceedings of the IEEE International Conference on Robotics and Automation*, volume 2, pages 1436–1443, April 2004.
42. DoD. Office of the secretary of defense unmanned systems roadmap (2007–2032), department of defense, washington, dc. 2007.
43. W. Dong, K. Zhou, Q. Fu, and Y. Duan. *Bio-Inspired Computing - Theories and Applications: 10th International Conference, BIC-TA 2015 Hefei, China, September 25–28, 2015, Proceedings*, chapter Adaptive Neighborhood Search's DGSO Applied to Travelling Saleman Problem, pages 125–137. Springer Berlin Heidelberg, Berlin, Heidelberg, 2015.
44. M. Dorigo. *Optimization, Learning and Natural Algorithms*. PhD thesis, Politecnico di Milano, Italy, 1992.
45. M. Dorigo, M. Birattari, and T. Stutzle. Ant colony optimization. *IEEE Computational Intelligence Magazine*, 1(4):28–39, 2006.
46. M. Dorigo and L.M. Gambardella. Ant colony system: a cooperative learning approach to the traveling salesman problem. *IEEE Transactions on Evolutionary Computation*, 1(1):53–66, 1997.
47. M. Dorigo, V. Maniezzo, and A. Colorni. Ant system: optimization by a colony of cooperating agents. *IEEE Transactions on Systems, Man, and Cybernetics, Part B (Cybernetics)*, 26(1):29–41, 1996.
48. M. Dorigo and T. Stutzle.*Ant Colony Optimization*. MIT Press, 2004.
49. J. Dréo and P. Siarry. Continuous interacting ant colony algorithm based on dense heterarchy. *Future Generation Computer Systems*, 20(5):841–856, 2004.
50. M. Du, X. Lei, and Z. Wu. A simplified glowworm swarm optimization algorithm. In *Proceedings of the IEEE Congress on Evolutionary Computation*, pages 2861–2868, July 2014.
51. R. Dutta, R. Ganguli, and V. Mani. Swarm intelligence algorithms for integrated optimization of piezoelectric actuator and sensor placement and feedback gains. *Smart Materials and Structures*, 20(10):105018, 2011.
52. H. El-Zabadani, A. Helal, and H. Yang. A mobile sensor platform approach to sensing and mapping pervasive spaces and their contents. In *Proceedings of the International Conference on New Technologies of Distributed Systems*, pages 4–8, 2007.
53. J.A. Farrell, S. Pang, W. Li, and R. Arrieta. Chemical plume tracing experimental results with a remus auv. In *Proceedings of the OCEANS Conference*, 2003.
54. N.R. Franks. Teams in social insects: group retrieval of prey by army ants (eciton burchelli, hymenoptera: Formicidae). *Behavioral Ecology and Sociobiology*, 18(6):425–429, 1986.
55. N.R. Franks, N. Gomez, S. Goss, and J.L. Deneubourg. The blind leading the blind in army ant raid patterns: testing a model of self-organization (hymenoptera: Formicidae). *Journal of Insect Behavior*, 4(5):583–607, 1991.
56. J.W. Fronczek and N.R. Prasad. Bio-inspired sensor swarms to detect leaks in pressurized systems. In *Proceedings of the IEEE International Conference on Systems, Man, and Cybernetics*, 2005.
57. L.A. Fuiman and A.E. Magurran. Development of predator defences in fishes. *Reviews in Fish Biology and Fisheries*, 4(2):145–183, 1994.
58. L.M. Gambardella and M. Dorigo. Ant-q: a reinforcement learning approach to the traveling salesman problem. In *Proceedings of the International Conference on Machine Learning*, pages 252–260, 2014.
59. S.K. Gazda, R.C. Connor, R.K. Edgar, and F. Cox. A division of labour with role specialization in group-hunting bottlenose dolphins (tursiops truncatus) off cedar key, florida. *Proceedings of the Royal Society of London B: Biological Sciences*, 272(1559):135–140, 2005.
60. V. Gazi and K.M. Passino. Stability analysis of swarms. *IEEE Transactions on Automatic Control*, 2003.

61. V. Gazi and K.M. Passino. Stability analysis of social foraging swarms. *IEEE Transactions on Systems, Man, and Cybernetics, Part B: Cybernetics*, 34(1):539–557, 2004.

62. B.P. Gerkey and R.T. Vaughan. Really reusable robot code and the player/stage project. *Software Engineering for Experimental Robotics, Springer Tracts on Advanced Robotics*, 2006.

63. M.R. Ghasemi and A. Ehsani. Multi-objective optimisation of composite laminates under heat and moisture effects using a hybrid neuro-ga algorithm. *International Journal of Civil, Environmental, Structural, Construction and Architectural Engineering*, 1(1):2–7, 2007.

64. D.E. Goldberg and J. Richardson. Genetic algorithms with sharing for multimodal function optimization. In *Proceedings of the International Conference on Genetic Algorithms and Their Applications*, pages 41–49, 1987.

65. Q. Gong, Y. Zhou, and Q. Luo. Hybrid artificial glowworm swarm optimization algorithm for solving multi-dimensional knapsack problem. *Procedia Engineering*, 15:2880–2884, 2011. CEIS 2011.

66. F.W. Grasso and J. Atema. Integration of flow and chemical sensing for guidance of autonomous marine robots in turbulent flows. *Environmental Fluid Mechanics*, 2(1):95–114, 2002.

67. A.O. Griewank. Generalized descent for global optimization. *Journal of Optimization Theory and Applications*, 34(1):11–39, 1981.

68. D.R. Griffin. *Bird migration*. Dover Publications, 1974.

69. J. Gu and K. Wen. Glowworm swarm optimization algorithm with quantum-behaved properties. In *Proceedings of the International Conference on Natural Computation*, pages 430–436, Aug 2014.

70. E.W. Gudger. A historical note on the synchronous flashing of fireflies. *Science*, 50(1286):188–190, 1919.

71. C.T. Hardin, X. Cui, R.K. Ragade, J.H. Graham, and A.S. Elmaghraby. A modified particle swarm algorithm for robotic mapping of hazardous environments. In *Proceedings of the World Automation Congress*, volume 17, pages 31–36, June 2004.

72. G.R. Harik. Finding multimodal solutions using restricted tournament selection. In *Proceedings of the International Conference on Genetic Algorithms*, pages 24–31, 1995.

73. A.T. Hayes, A. Martinoli, and R.M. Goodman. Swarm robotic odor localization: Off-line optimization and validation with real robots. *Robotica*, 21(4):427–441, 2003.

74. D. He and H. Zhu. Glowworm swarm optimization algorithm based on multi-population. In *Proceedings of the International Conference on Natural Computation*, volume 5, pages 2624–2627, Aug 2010.

75. L. He, X. Tong, and S. Huang. A glowworm swarm optimization algorithm with improved movement rule. In *Proceedings of the International Conference on Intelligent Networks and Intelligent Systems*, pages 109–112, Nov 2012.

76. L. He, X. Tong, and S. Huang. Glowworm swarm optimization algorithm based on hierarchical multi-subgroups. *Journal of Information and Computational Science*, 10(4):1245–1251, 2013.

77. L. He, X. Tong, S. Huang, and Q. Wang. Glowworm swarm optimization algorithm with improved movement pattern. In *Proceedings of the International Conference on Intelligent Networks and Intelligent Systems*, pages 43–46, Nov 2013.

78. J. Hereford and M. Siebold. Multi-robot search using a physically-embedded particle swarm optimization. *International Journal of Computational Intelligence Research*, 4(2), 2008.

79. S.C.J.J.L. Higdon. Induced drag of a bird flock. *The American Naturalist*, 112(986):727–744, 1978.

80. D.J. Hoare, I.D. Couzin, J.-G.J. Godin, and J. Krause. Context-dependent group size choice in fish. *Animal Behaviour*, 67(1):155–164, 2004.

81. D.S. Hochbaum and D.B. Shmoys. A best possible heuristic for the k-center problem. *Mathematics of Operations Research*, 10(2):180–184, 1985.

82. R. Hooke and T.A. Jeeves. " direct search" solution of numerical and statistical problems. *Journal of the ACM*, 8(2):212–229, 1961.

83. J. Horn, D.E. Goldberg, and K. Deb. Implicit niching in a learning classifier system: Nature's way. *Evolutionary Computation*, 2(1):37–66, 1994.

84. K. Huang and Y.Q. Zhou. *Bio-Inspired Computing and Applications: 7th International Conference on Intelligent Computing, ICIC 2011, Zhengzhou, China, August 11–14. 2011, Revised Selected Papers*, chapter A Novel Chaos Glowworm Swarm Optimization Algorithm for Optimization Functions, pages 426–434. Springer Berlin Heidelberg, Berlin, Heidelberg, 2012.
85. Z. Huang and Y. Zhou. Using glowworm swarm optimization algorithm for clustering analysis. *Journal of Convergence Information Technology*, 6(2):78–85, 2011.
86. L. Ingber and B. Rosen. Genetic algorithms and very fast simulated reannealing: A comparison. *Math. Comput. Model.*, 16(11):87–100, 1992.
87. H. Jabbar, W.A. Lughmani, S. Yoo, and T. Jeong. Ubiquitous factory: An automation of nuclear power plant. In *Proceedings of the International Conference on Ubiquitous Information Technologies and Applications*, pages 1266–1273, 2007.
88. M.V. Jakuba. *Stochastic Mapping for Chemical Plume Source Localization with Application to Autonomous Hydrothermal Vent Discovery*. PhD thesis, Massachusetts Institute of Technology and Woods Hole Oceanographic Institution, 2007.
89. D.N. Jayakumar and P. Venkatesh. Glowworm swarm optimization algorithm with topsis for solving multiple objective environmental economic dispatch problem. *Applied Soft Computing*, 23(0):375–386, 2014.
90. H. Jiang and X. Tang. Polarimetric mimo radar target detection based on glowworm swarm optimization algorithm. In *Proceedings of the IEEE International Conference on Acoustics, Speech and Signal Processing*, pages 805–809, May 2014.
91. Z.H. Jiang. *Fixed Point Theory*. Beijing: Science Press, 1979.
92. D. Jinxin and Q. Minyong. New clustering algorithm based on particle swarm optimization and simulated annealing. *Computer Engineering and Applications*, 45(35):139–141, 2009.
93. Y.-J. Yang K. Kao and K.-T. Sang. A glowworm algorithm for solving data clustering problems. In *Proceedings of the International Conference on Accounting and Information Technology*. Taipei, Taiwan, August 2008.
94. Z. Kang, X. He, and J. Liu. X-ray pulsars time delay estimation using gso-based bispectral feature points. *Optik-International Journal for Light and Electron Optics*, 127(12):5050–5054, 2016.
95. D. Karaboga. An idea based on honey bee swarm for numerical optimization. *Tech. Report-TR06, Erciyes University, Engineering Faculty, Computer Engineering Department*, 2005.
96. P. Kavipriya. Ovsf dynamic code allocation quality based glowworm swarm optimization approach in wcdm. *Indian Journal of Science and Technology*, 8(12), 2015.
97. J. Kennedy. Stereotyping: improving particle swarm performance with cluster analysis. In *Proceedings of the Congress on Evolutionary Computation*, pages 1507–1512. San Diego, California, July 2000.
98. J. Kennedy and R. Eberhart. Particle swarm optimization. In *Proceedings of IEEE International Conference on Neural Networks*, pages 1942–1948. Perth, WA, Australia, 1995.
99. J. Kennedy and R.C. Eberhart. A discrete binary version of the particle swarm algorithm. In *Proceedings of the IEEE International Conference on Systems, Man, and Cybernetics*, 1997.
100. K. Khan and A. Sahai. A glowworm optimization method for the design of web services. *International Journal of Intelligent Systems and Applications*, 4(10):89, 2012.
101. H.-C. Kim, Y.J. Jung, B.-S. Kim, K.H. Ryu, S. Eun, and K.-O. Kim. A framework for atmospheric environment monitoring with ubiquitous sensor network. In *Proceedings of the International Conference on Ubiquitous Information Technologies and Applications*, pages 1240–1248, 2007.
102. T. Kohonen. The handbook of brain theory and neural networks. chapter Learning Vector Quantization, pages 537–540. MIT Press, Cambridge, MA, USA, 1998.
103. K.N. Krishnanand. *Glowworm Swarm Optimization: A Multimodal Function Optimization Paradigm with Applications to Multiple Signal Source Localization Tasks*. PhD thesis, Indian Institute of Science, Bangalore, 2007.
104. K.N. Krishnanand, P. Amruth, M.H. Guruprasad, S.V. Bidargaddi, and D. Ghose. Glowworm-inspired robot swarm for simultaneous taxis towards multiple radiation sources. In *Proceedings of the IEEE International Conference on Robotics and Automation*, pages 958–963. Orlando, Florida, May 2006.

105. K.N. Krishnanand and D. Ghose. Detection of multiple source locations using a glowworm metaphor with applications to collective robotics. In *Proceedings of the IEEE Swarm Intelligence Symposium*, pages 84–91. Pasedena, California, June 2005.

106. K.N. Krishnanand and D. Ghose. Multimodal function optimization using a glowworm metaphor with applications to collective robotics. In *Proceedings of the Indian International Conference on Artificial Intelligence*, pages 328–346. Pune, India, 2005.

107. K.N. Krishnanand and D. Ghose. Glowworm swarm based optimization algorithm for multimodal functions with collective robotics applications. *Multiagent and Grid Systems*, 2(3):209–222, 2006.

108. K.N. Krishnanand and D. Ghose. *Design and Control of Intelligent Robotic Systems*, chapter A Glowworm Swarm Optimization Based Multi-robot System for Signal Source Localization, pages 49–68. Springer Berlin Heidelberg, Berlin, Heidelberg, 2009.

109. K.N. Krishnanand and D. Ghose. Glowworm swarm optimization for simultaneous capture of multiple local optima of multimodal functions. *Swarm Intelligence*, 3(2):87–124, 2009.

110. K.N. Krishnanand and D. Ghose. *Handbook of Swarm Intelligence: Concepts, Principles and Applications*, chapter Glowworm swarm optimization for multimodal search spaces, pages 451–467. Springer Berlin Heidelberg, Berlin, Heidelberg, 2011.

111. K.N. Krishnanand and Debasish Ghose. Formations of minimalist mobile robots using local-templates and spatially distributed interactions. *Robotics and Autonomous Systems*, 53(3–4):194–213, 2005.

112. K.N. Krishnanand, A. Puttappa, G.M. Hegde, S.V. Bidargaddi, and D. Ghose. *Ant Colony Optimization and Swarm Intelligence: 5th International Workshop, ANTS 2006, Brussels, Belgium, September 4–7, 2006*. Proceedings, chapter Rendezvous of Glowworm-Inspired Robot Swarms at Multiple Source Locations: A Sound Source Based Real-Robot Implementation, pages 259–269. Springer Berlin Heidelberg, Berlin, Heidelberg, 2006.

113. K.N. Krishnanand and D.Ghose. Glowworm swarm optimisation: a new method for optimising multi-modal functions. *International Journal of Computational Intelligence Studies*, 1(1): 93–119, 2009.

114. A.-C. Lee and S.-T. Chen. Collocated sensor/actuator positioning and feedback design in the control of flexible structure system. *Journal of Vibration and Acoustics*, 116(2):146–154, 1994.

115. S.M. Lewis and C.K. Cratsley. Flash signal evolution, mate choice, and predation in fireflies. *Annual Review of Entomology*, 53:293–321, 2008.

116. M. Li, X. Wang, Y. Gong, Y. Liu, and C. Jiang. Binary glowworm swarm optimization for unit commitment. *Journal of Modern Power Systems and Clean Energy*, 2(4):357–365, 2014.

117. X. Li. *Genetic and Evolutionary Computation - GECCO 2004: Genetic and Evolutionary Computation Conference, Seattle, WA, USA, June 26–30, 2004. Proceedings, Part I*, chapter Adaptively choosing neighbourhood bests using species in a particle swarm optimizer for multimodal function optimization, pages 105–116. Springer Berlin Heidelberg, Berlin, Heidelberg, 2004.

118. Z. Li and X. Huang. Glowworm swarm optimization and its application to blind signal separation. *Mathematical Problems in Engineering*, 2016, 2016.

119. T. Liang, C. Qiang, N. Yurong, Y. Suhua, and S. Guofa. Adaptive control of mechanical servo system with glowworm swarm friction identification. In *Proceedings of the Chinese Control Conference*, pages 3132–3138, July 2015.

120. L. Liangdong Qu, D. He, and J. Wu. Hybrid co-evolutionary glowworm swarm optimization algorithm for fixed point equation. *Journal of Information and Computational Science*, 2011.

121. W.-H. Liao, Y. Kao, and Y.-S. Li. A sensor deployment approach using glowworm swarm optimization algorithm in wireless sensor networks. *Expert Systems with Applications*, 38(10):12180–12188, 2011.

122. P.B.S. Lissaman and C.A. Shollenberger. Formation flight of birds. *Science*, 168(3934):1003–1005, 1970.

123. J. Liu, Y. Zhou, K. Huang, Z. Ouyang, and Y. Wang. A glowworm swarm optimization algorithm based on definite updating search domains. *Journal of Computational Information Systems*, 7(10):3698–3705, 2011.

124. J.E. Lloyd. Bioluminescent communication in insects. *Annual Review of Entomology*, 16(1):97–122, 1971.
125. R.G. Loewy. Recent developments in smart structures with aeronautical applications. *Smart Materials and Structures*, 6(5):R11, 1997.
126. M. Løvbjerg, T.K. Rasmussen, and T. Krink. Hybrid particle swarm optimizer with breeding and subpopulations. In *Proceedings of the Genetic and Evolutionary Computation Conference*. San Francisco, California, July 2001.
127. C. Lytridis, G.S. Virk, and E.E. Kadar. *Climbing and Walking Robots: Proceedings of the 7th International Conference CLAWAR 2004*, chapter Co-operative smell-based navigation for mobile robots, pages 1107–1117. Springer Berlin Heidelberg, Berlin, Heidelberg, 2005.
128. J. MacQueen. Some methods for classification and analysis of multivariate observations. In *Proceedings of the Berkeley Symposium on Mathematical Statistics and Probability*, pages 281–297, 1967.
129. R. Mageshvaran and T. Jayabarathi. GSO based optimization of steady state load shedding in power systems to mitigate blackout during generation contingencies. *Ain Shams Engineering Journal*, 6(1):145–160, 2015.
130. A.E. Magurran and T.J. Pitcher. Provenance, shoal size and the sociobiology of predator-evasion behaviour in minnow shoals. *Proceedings of the Royal Society of London B: Biological Sciences*, 229(1257):439–465, 1987.
131. S.W. Mahfoud. Crowding and preselection revisited. In *Proceedings of the International Conference on Parallel Problem Solving from Nature*, pages 28–30. Brussels, Belgium, 1992.
132. S.W. Mahfoud. *Niching Methods for Genetic Algorithms*. PhD thesis, University of Illinois at Urbana-Champaign, 1995.
133. K.K. Mandal and N. Chakraborty. Effect of control parameters on differential evolution based combined economic emission dispatch with valve-point loading and transmission loss. *International Journal of Emerging Electric Power Systems*, 9(4), 2008.
134. V. Maniezzo. Exact and approximate nondeterministic tree-search procedures for the quadratic assignment problem. *INFORMS Journal on Computing*, 11(4):358–369, 1999.
135. P. Manimaran and V. Selladurai. Glowworm swarm optimisation algorithm for nonlinear fixed charge transportation problem in a single stage supply chain network. *International Journal of Logistics Economics and Globalisation*, 6(1):42–55, 2014.
136. R. Manoharan and A.K. Jeevanantham. Stress and load-displacement analysis of fiber reinforced composite laminates with a circular hole under compressive load. *ARPN Journal of Engineering and Applied Sciences*, 6(4):64–74, 2011.
137. M. Marinaki and Y. Marinakis. A glowworm swarm optimization algorithm for the vehicle routing problem with stochastic demands. *Expert Systems with Applications*, 46:145–163, 2016.
138. L. Marques, U. Nunes, and A.T. de Almeida. Particle swarm-based olfactory guided search. *Autonomous Robots*, 20(3):277–287, 2006.
139. M.J. Mataric. Learning in behavior-based multi-robot systems: policies, models, and other agents. *Cognitive Systems Research*, 2(1):81–93, 2001.
140. K. McGill and S. Taylor. Comparing swarm algorithms for large scale multi-source localization. In *Proceedings of the IEEE International Conference on Technologies for Practical Robot Applications*, pages 48–54, Nov 2009.
141. K. McGill and S. Taylor. Comparing swarm algorithms for multi-source localization. In *Proceedings of the IEEE International Workshop on Safety, Security, and Rescue Robotics*, pages 1–7, Nov 2009.
142. K. McGill and S. Taylor. Diffuse algorithm for robotic multi-source localization. In *Proceedings of the IEEE International Conference on Technologies for Practical Robot Applications*, pages 121–126, April 2011.
143. K. McGill and S. Taylor. Robot algorithms for localization of multiple emission sources. *ACM Computing Surveys*, 43(3):15:1–15:25, April 2011.
144. K.C. Meher, R.K. Swain, and C.K. Chanda. Modified gso for combined economic emission load dispatch with valve-point effects. In *Proceedings of the International Conference on Advances in Electronics, Computers and Communications*, pages 1–6, Oct 2014.

145. K.C. Meher, R.K. Swain, and C.K. Chanda. Dynamic economic dispatch using glowworm swarm optimization technique. *International Journal of Electrical, Computer, Energetic, Electronic and Communication Engineering*, 9(9):1039–1044, 2016.

146. O.J. Mengshoel and D.E. Goldberg. Probabilistic crowding: deterministic crowding with probabilistic replacement. In *Proceedings of the Genetic and Evolutionary Computation Conference*, page 409, 1999.

147. P.P. Menon and D. Ghose. Simultaneous source localization and boundary mapping for contaminants. In *Proceedings of the American Control Conference (ACC)*, pages 4174–4179, June 2012.

148. P.P. Menon and D. Ghose. Boundary mapping of 3-dimensional regions. In *Proceedings of the American Control Conference*, pages 2984–2989, June 2013.

149. V.B. Meyer-Rochow. Glowworms: a review of arachnocampa spp. and kin. *Luminescence*, 22(3):251–265, 2007.

150. Z. Michalewicz. *Genetic algorithms + data structures = evolution programs*. Springer Science and Business Media, 2013.

151. B.L. Miller and M.J. Shaw. Genetic algorithms with dynamic niche sharing for multimodal function optimization. In *Proceedings of the IEEE International Conference on Evolutionary Computation*, pages 786–791, May 1996.

152. Y. Mo, F. Liu, and Y. Ma. *Intelligent Computing Theories: 9th International Conference, ICIC 2013, Nanning, China, July 28–31, 2013. Proceedings*, chapter Application of GSO Algorithm to the Parameter Estimation of Option Pricing Model, pages 199–206. Springer Berlin Heidelberg, Berlin, Heidelberg, 2013.

153. A. Moiseff and J. Copeland. Firefly synchrony: a behavioral strategy to minimize visual clutter. *Science*, 329(5988):181–181, 2010.

154. F. Mondada, M. Bonani, X. Raemy, J. Pugh, C. Cianci, A. Klaptocz, S. Magnenat, J.-C. Zufferey, D. Floreano, and A. Martinoli. The e-puck, a robot designed for education in engineering. In *Proceedings of the Conference on Autonomous Robot Systems and Competitions*, pages 59–65, 2009.

155. M.J. Morgan. The influence of hunger, shoal size and predator presence on foraging in bluntnose minnows. *Animal Behaviour*, 36(5):1317–1322, 1988.

156. E.S. Morse. Fireflies flashing in unison. *Science*, 48(1230):92–93, 1918.

157. H. Mühlenbein, M. Schomisch, and J. Born. The parallel genetic algorithm as function optimizer. *Parallel Computing*, 17(6):619–632, 1991.

158. M.N. Mukherjee. *Elements of Metric Spaces*. Academic Publishers, 2010.

159. S.D. Muller, J. Marchetto, S. Airaghi, and P. Kournoutsakos. Optimization based on bacterial chemotaxis. *IEEE Transactions on Evolutionary Computation*, 6(1):16–29, Feb 2002.

160. J. Murlis and C.D. Jones. Fine-scale structure of odour plumes in relation to insect orientation to distant pheromone and other attractant sources. *Physiological Entomology*, 6(1):71–86, 1981.

161. W. Naeem, R. Sutton, and J. Chudley. Chemical plume tracing and odour source localisation by autonomous vehicles. *The Journal of Navigation*, 60(2):173–190, 2007.

162. J.A. Nelder and R. Mead. A simplex method for function minimization. *The Computer Journal*, 7(4):308–313, 1965.

163. P. Oramus. Improvements to glowworm swarm optimization algorithm. *Computer Science*, 11:7–20, 2010.

164. P. Oramus. Further improvement of the glowworm swarm optimization algorithm by adding a conservation of agent move direction. *Bio-Algorithms and Med-Systems*, 7(4), 2011.

165. A. Ouyang, L. Liu, G. Yue, X. Zhou, and K. Li. Bfgs-gso for global optimization problems. *Journal of Computers*, 9(4), 2014.

166. C. Packer and L. Ruttan. The evolution of cooperative hunting. *American Naturalist*, 132(2):159–198, 1988.

167. K.E. Parsopoulos, V.P. Plagianakos, G.D. Magoulas, and M.N. Vrahatis. A clearing procedure as a niching method for genetic algorithms. In *Proceedings of the Particle Swarm Optimization Workshop*. Indianapolis, Indiana, April 2001.

168. K.E. Parsopoulos and M.N. Vrahatis. On the computation of all global minimizers through particle swarm optimization. *IEEE Transactions on Evolutionary Computation*, 8(3):211–224, June 2004.
169. B.L. Partridge and T.J. Pitcher. The sensory basis of fish schools: Relative roles of lateral line and vision. *Journal of comparative physiology*, 135(4):315–325, 1980.
170. M.E.H. Pedersen. *Tuning and Simplifying Heuristical Optimization*. PhD thesis, University of Southampton, 2010.
171. D. Pengzhen, T. Zhenmin, and S. Yan. A quantum glowworm swarm optimization algorithm based on chaotic sequence. *International Journal of Control and Automation*, 7(9):165–178, 2014.
172. A. Petrowski. A clearing procedure as a niching method for genetic algorithms. In *Proceedings of the IEEE International Conference on Evolutionary Computation*, pages 798–803, May 1996.
173. D.T. Pham, A. Ghanbarzadeh, E. Koc, S. Otri, S. Rahim, and M. Zaidi. The bees algorithm-a novel tool for complex optimisation. In *Proceedings of the Virtual International Conference on Intelligent Production Machines and Systems*, pages 454–459. Elsevier, July 2006.
174. S. Poduri and G.S. Sukhatme. Constrained coverage for mobile sensor networks. In *Proceedings of the IEEE International Conference on Robotics and Automation*, volume 1, pages 165–171 Vol.1, April 2004.
175. R. Poli. Analysis of the publications on the applications of particle swarm optimisation. *Journal of Artificial Evolution and Applications*, 2008:3:1–3:10, 2008.
176. R. Poli, J. Kennedy, and T. Blackwell. Particle swarm optimization. *Swarm Intelligence*, 1(1):33–57, 2007.
177. H. Pubo and W. Yu. Nonlinear blind source separation algorithm using glowworm swarm optimization with baffle effect. *Telecommunications Science*, 31(9):2015213, 2015.
178. J. Pugh and A. Martinoli. Multi-robot learning with particle swarm optimization. In *Proceedings of the International Joint Conference on Autonomous Agents and Multiagent Systems*, AAMAS '06, pages 441–448, New York, NY, USA, 2006. ACM.
179. J. Pugh and A. Martinoli. Inspiring and modeling multi-robot search with particle swarm optimization. In *Proceedings of the IEEE Swarm Intelligence Symposium*, pages 332–339, April 2007.
180. K. Pushpalatha and V.S. Ananthanarayana. A new glowworm swarm optimization based clustering algorithm for multimedia documents. In *Proceedings of the IEEE International Symposium on Multimedia*, pages 262–265, Dec 2015.
181. B. Raman and S. Subramani. An efficient specific update search domain based glowworm swarm optimization for test case prioritization. *International Arab Journal of Information Technology*, 12, 2015.
182. A. Ray and D. De. An energy efficient sensor movement approach using multi-parameter reverse glowworm swarm optimization algorithm in mobile wireless sensor network. *Simulation Modelling Practice and Theory*, 62:117–136, 2016.
183. S.S. Reddy and C.S. Rathnam. Optimal power flow using glowworm swarm optimization. *International Journal of Electrical Power and Energy Systems*, 80:128–139, 2016.
184. S.Y. Reutskiy and C.S. Chen. Approximation of multivariate functions and evaluation of particular solutions using chebyshev polynomial and trigonometric basis functions. *International Journal for Numerical Methods in Engineering*, 67(13):1811–1829, 2006.
185. G. Sandini, G. Lucarini, and M. Varoli. Gradient driven self-organizing systems. In *Proceedings of the IEEE/RSJ International Conference on Intelligent Robots and Systems*, pages 429–432, Jul 1993.
186. P. Scerri, T. Von Gonten, G. Fudge, S. Owens, and K. Sycara. Transitioning multiagent technology to uav applications. In *Proceedings of the International Joint Conference on Autonomous Agents and Multiagent Systems: Industrial Track*, AAMAS '08, pages 89–96, Richland, SC, 2008. Estoril, Portugal, International Foundation for Autonomous Agents and Multiagent Systems.

187. H-P. Schwefel. *Numerical optimization of computer models*. John Wiley and Sons, Inc., New York, NY, USA, 1981.

188. G.F. Scimemi and S. Rizzo. Optimization of laminated composites plates using glowowrm algorithm. In *Proceedings of the GIMC Conference, Siracusa*, 2010.

189. D.T. Seeley and C.S. Buhrman. Group decision making in swarms of honey bees. *Behavioral Ecology and Sociobiology*, 45(1):19–31, 1999.

190. T.D. Seeley. Division of labor between scouts and recruits in honeybee foraging. *Behavioral Ecology and Sociobiology*, 12(3):253–259, 1983.

191. M. Senanayake, I. Senthooran, J.C. Barca, H. Chung, J. Kamruzzaman, and M. Murshed. Search and tracking algorithms for swarms of robots: A survey. *Robotics and Autonomous Systems*, 75, Part B:422–434, 2016.

192. G. Seront and L. Gambardella. Results of the first international contest on evolutionary optimisation (1st iceo). 1996.

193. P.S. Shelokar, V.K. Jayaraman, and B.D. Kulkarni. An ant colony approach for clustering. *Analytica Chimica Acta*, 509(2):187–195, 2004.

194. O. Shimomura. *Bioluminescence: Chemical Principles and Methods*. World Scientific, 2012.

195. R. Sikora and M. Shaw. A double-layered learning approach to acquiring rules for classification: Integrating genetic algorithms with similarity-based learning. *ORSA Journal on Computing*, 6(2):174–187, 1994.

196. A. Singh and K. Deep. New variants of glowworm swarm optimization based on step size. *International Journal of System Assurance Engineering and Management*, 6(3):286–296, 2015.

197. A. Singh and K. Deep. *Proceedings of the International Conference on Soft Computing for Problem Solving*, chapter How improvements in glowworm swarm optimization can solve real-life problems, pages 279–291. Springer India, New Delhi, 2015.

198. H.M. Smith. Synchronous flashing of fireflies. *Science*, 82(2120):151–152, 1935.

199. K. Socha and M. Dorigo. Ant colony optimization for continuous domains. *European Journal of Operational Research*, 185(3):1155–1173, 2008.

200. S.H. Strogatz. Exploring complex networks. *Nature*, 410(6825):268–276, 2001.

201. T. Stützle and H.H. Hoos. Max-min ant system. *Future Generation Computer Systems*, 16(8):889–914, 2000.

202. Z. Tang and Y. Zhou. A glowworm swarm optimization algorithm for uninhabited combat air vehicle path planning. *Journal of Intelligent Systems*, 24(1):69–83, 2015.

203. Z. Tang, Y. Zhou, and X. Chen. An improved glowworm swarm optimization algorithm based on parallel hybrid mutation. In D.-S. Huang, K.-H. Jo, Y.-Q. Zhou, and K. Han, editors, *Intelligent Computing Theories and Technology, volume 7996 of Lecture Notes in Computer Science*, pages 198–206. Springer Berlin Heidelberg, 2013.

204. E. Thiéemard. Economic generation of low-discrepancy sequences with a b-ary gray code. *Department of Mathematics, Ecole Polytechnique Fédérale de Lausanne, Lausanne, Switzerland*, 1998.

205. J. Thomas. Odor source localization using swarm robotics. Master's thesis, Indian Institute of Science, Bangalore, 2008.

206. J. Thomas and D. Ghose. Strategies for locating multiple odor sources using glowworm swarm optimization. In *Indian International Conference on Artificial Intelligence*, pages 842–861, 2009.

207. A. Torn and A. Zilinskas. *Global Optimization*. Springer-Verlag New York, Inc., New York, NY, USA, 1989.

208. F. Valdez and P. Melin. Parallel evolutionary computing using a cluster for mathematical function optimization. In *NAFIPS 2007 - 2007 Annual Meeting of the North American Fuzzy Information Processing Society*, pages 598–603, June 2007.

209. T. Vicsek, A. Czirók, E. Ben-Jacob, I. Cohen, and O. Shochet. Novel type of phase transition in a system of self-driven particles. *Physical Review Letters*, 75:1226–1229, 1995.

210. K. Von Frisch. *The Dance Language and Orientation of Bees*. Harvard University Press, 1967.

211. J. Wang, Y. Cao, B. Li, S. Lee, and J.-U. Kim. A glowworm swarm optimization based clustering algorithm with mobile sink support for wireless sensor networks. *Journal of Internet Technology*, 16(5):825–832, 2015.

212. J. Wang, L. Dong-Xu, and J. Jiang. Integrated control of thermally induced vibration and quasi-static deformation of space truss. *Journal of Dynamic Systems, Measurement, and Control*, 2016.

213. J.-S. Wang, S. Han, and N.-N. Shen. Improved gso optimized esn soft-sensor model of flotation process based on multisource heterogeneous information fusion. *The Scientific World Journal*, 2014, 2014.

214. L. Wang, L. Zhao, and H. Yan. Application of gso for load allocation between hydropower units and its model analysis based on multi-objective. *Journal of Computers*, 7(5), 2012.

215. Y.-C. Wang, C.-C. Hu, and Y.-C. Tseng. Efficient deployment algorithms for ensuring coverage and connectivity of wireless sensor networks. In *Proceedings of the International Conference on Wireless Internet*, pages 114–121, July 2005.

216. M. Weiser, R. Gold, and J.S. Brown. The origins of ubiquitous computing research at parc in the late 1980s. *IBM systems journal*, 38(4):693, 1999.

217. E.A. Widder. Bioluminescence in the ocean: origins of biological, chemical, and ecological diversity. *Science*, 328(5979):704–708, 2010.

218. E.A. Widder. Glowing life in an underwater world. In *TED Talk, Technology, Entertainment, and Design Conference*, 2010.

219. B. Wu, C. Qian, W. Ni, and S. Fan. The improvement of glowworm swarm optimization for continuous optimization problems. *Expert Systems with Applications*, 39(7):6335–6342, 2012.

220. B. Xing. Trends in Ambient Intelligent Systems: The Role of Computational Intelligence, chapter Smart Robot Control via Novel Computational Intelligence Methods for Ambient Assisted Living, pages 29–55. Springer International Publishing, Cham, 2016.

221. B. Xing. *Trends in Ambient Intelligent Systems: The Role of Computational Intelligence*, chapter An Investigation of the Use of Innovative Biology-Based Computational Intelligence in Ubiquitous Robotics Systems: Data Mining Perspective, pages 139–172. Springer International Publishing, Cham, 2016.

222. X-S Yang. *Proceedings of the 5th International Symposium on Stochastic Algorithms: Foundations and Applications, Sapporo, Japan*, chapter Firefly algorithms for multimodal optimization, pages 169–178. Springer Berlin Heidelberg, Berlin, Heidelberg, October 2009.

223. Y. Yang and Y. Zhou. *Proceedings of the International Conference on Intelligent Computing and Information Science, Chongqing, China, January*, chapter Glowworm Swarm Optimization Algorithm for Solving Numerical Integral, pages 389–394. Springer Berlin Heidelberg, Berlin, Heidelberg, 2011.

224. Y. Yang, Y. Zhou, and Q. Gong. Hybrid artificial glowworm swarm optimization algorithm for solving system of nonlinear equations. *Journal of Computational Information Systems*, 6(10):3431–3438, 2010.

225. V. Yepes, J.V. Marti, and T. Garcia-Segura. Cost and co_2 emission optimization of precast-prestressed concrete u-beam road bridges by a hybrid glowworm swarm algorithm. *Automation in Construction*, 49, Part A:123–134, 2015.

226. Z. Yu and X. Yang. Full glowworm swarm optimization algorithm for whole-set orders scheduling in single machine. *The Scientific World Journal*, 2013, 2013.

227. D. Zarzhitsky, D.F. Spears, and W.M. Spears. Swarms for chemical plume tracing. In *Proceedings of the IEEE Swarm Intelligence Symposium*, pages 249–256, June 2005.

228. H. Zhang and J.C. Hou. Maintaining sensing coverage and connectivity in large sensor networks. *Ad Hoc and Sensor Wireless Networks*, 1(1–2):89–124, 2005.

229. H. Zhang, B. Lennox, H. Zhang, P.R. Goulding, and A.Y.T. Leung. A float-encoded genetic algorithm technique for integrated optimization of piezoelectric actuator and sensor placement and feedback gains. *Smart Materials and Structures*, 9(4):552, 2000.

230. H. Zhang, Y. Liu, N.N. Xiong, M. Imran, and A.V. Vasilakos. An improved similarity based adaptive step size glowworm algorithm. *Journal of Internet Technology*, 16(5):905–914, 2015.

231. J.M. Zhang. A new proof of convergence order for an iterative method and generalization. *College Mathematics*, 23(6):135–139, 2007.

232. Y. Zhang, X. Ma, and Y. Miao. Localization of multiple odor sources using modified glowworm swarm optimization with collective robots. In *Proceedings of the Chinese Control Conference*, pages 1899–1904, July 2011.

233. Y. L. Zhang, X. P. Ma, Y. Gu, and Y. Z. Miao. A modified glowworm swarm optimization for multimodal functions. In *Proceedings of the Chinese Control and Decision Conference*, pages 2070–2075, May 2011.

234. G. Zhao, Y. Zhou, and Y. Wang. Using complex method guidance gso swarm algorithm for solving high dimensional function optimization problem. *Journal of Convergence Information Technology*, 6(11), 2011.

235. W. Zhao and D. Wei. Relevance vector machine combined with glowworm swarm optimization for cam relationship of kaplan turbine. *Advanced Science Letters*, 11(1):244–247, 2012.

236. X. Zhao, S. Chen, and Y. Sheng. Research on the problem of the shortest path based on the glowworm swarm optimization algorithm. In *Proceedings of the International Conference of Transportation Professionals*, 2015.

237. Y. Zhao, G. Karypis, and U. Fayyad. Hierarchical clustering algorithms for document datasets. *Data Mining and Knowledge Discovery*, 10(2):141–168, 2005.

238. G.-D. Zhou, T.-H. Yi, H. Zhang, and H.-N. Li. Optimal sensor placement under uncertainties using a nondirective movement glowworm swarm optimization algorithm. *Smart Structures and Systems*, 16(2):243–262, 2015.

239. Q. Zhou, Y. Zhou, and X. Chen. *Intelligent Computing Theories and Technology: 9th International Conference, ICIC 2013, Nanning, China, July 28–31, 2013. Proceedings*, chapter Cloud Model Glowworm Swarm Optimization Algorithm for Functions Optimization, pages 189–197. Springer Berlin Heidelberg, Berlin, Heidelberg, 2013.

240. Y. Zhou, J. Liu, and G. Zhao. Leader glowworm swarm optimization algorithm for solving nonlinear equations systems. *Electrical Review*, 88(1):101–106, 2012.

241. Y. Zhou, Q. Luo, and J. Liu. Glowworm swarm optimization for optimization dispatching system of public transit vehicles. *Journal of Theoretical and Applied Information Technology*, 52(2), 2013.

242. Y. Zhou, Z. Ouyang, J. Liu, and G. Sang. A novel k-means image clustering algorithm based on glowworm swarm optimization. *PRZEGLAD ELEKTROTECHNICZNY*, 88(8):266–270, 2012.

243. Y. Zhou, Y. Wang, S. He, and J. Wu. A novel double glowworm swarm co-evolution optimization algorithm based levy flights. *Applied Mathematics and Information Sciences*, 8(1L):355–361, 2014.

244. Y. Zhou, G. Zhou, and J. Zhang. A hybrid glowworm swarm optimization algorithm to solve constrained multimodal functions optimization. *Optimization*, 64(4):1057–1080, 2015.

245. Y. Zou and K. Chakrabarty. A swarm approach for emission sources localization. In *Proceedings of the Twenty-Second Annual Joint Conference of the IEEE Computer and Communications on Tools with Artificial Intelligence*, pages 1293–1303, March 2003.

Printed in the United States
By Bookmasters